Sophia Studies in Cross-cultural Philosophy of Traditions and Cultures

Volume 42

The Sophia Studies in Cross-cultural Philosophy of Traditions and Cultures focuses on the broader aspects of philosophy and traditional intellectual patterns of religion and cultures. The series encompasses global traditions, and critical treatments that draw from cognate disciplines, inclusive of feminist, postmodern, and postcolonial approaches. By global traditions we mean religions and cultures that go from Asia to the Middle East to Africa and the Americas, including indigenous traditions in places such as Oceania. Of course this does not leave out good and suitable work in Western traditions where the analytical or conceptual treatment engages Continental (European) or Cross-cultural traditions in addition to the Judeo-Christian tradition. The book series invites innovative scholarship that takes up newer challenges and makes original contributions to the field of knowledge in areas that have hitherto not received such dedicated treatment. For example, rather than rehearsing the same old Ontological Argument in the conventional way, the series would be interested in innovative ways of conceiving the erstwhile concerns while also bringing new sets of questions and responses, methodologically also from more imaginative and critical sources of thinking. Work going on in the forefront of the frontiers of science and religion beaconing a well-nuanced philosophical response that may even extend its boundaries beyond the confines of this debate in the West – e.g. from the perspective of the 'Third World' and the impact of this interface (or clash) on other cultures, their economy, sociality, and ecological challenges facing them – will be highly valued by readers of this series. All books to be published in this Series will be fully peer-reviewed before final acceptance.

Lenart Škof • Sashinungla
Sigridur Thorgeirsdottir
Editors

Elemental-Embodied Thinking for a New Era

Springer

Editors
Lenart Škof
Institute for Philosophical
and Religious Studies
Science and Research Centre Koper
Koper, Slovenia

Sashinungla
Department of Philosophy
Jadavpur University
Kolkata, India

Sigridur Thorgeirsdottir
Department of Philosophy
University of Iceland
Reykjavik, Iceland

Editor-in-Chief
Purushottama Bilimoria

ISSN 2211-1107 ISSN 2211-1115 (electronic)
Sophia Studies in Cross-cultural Philosophy of Traditions and Cultures
ISBN 978-3-031-42118-1 ISBN 978-3-031-42119-8 (eBook)
https://doi.org/10.1007/978-3-031-42119-8

Slovenian Research and Innovation Agency research grants: J7-1824, J6-50196, and P6-0279
The international research project Embodied Critical Thinking and the Erasmus+ Training Embodied Critical Thinking program.

This Springer imprint is published by the registered company Springer Nature Switzerland AG
The registered company address is: Gewerbestrasse 11, 6330 Cham, Switzerland

Paper in this product is recyclable.

Foreword

This is an important and timely book, notably for its rethinking on four fronts: engaging with the intellectual and environmental problems of the Anthropocene; theorizing embodied thinking, agency, and environmental practice; facilitating a broad take on sentient relational beings; and, finally, moving beyond the fragmented disciplines of the humanities. This is quite a challenge, and with profound implications. Significantly, a family resemblance of such ideas is currently floating around across disciplines and continents.

Philosopher Hannah Arendt famously argued, at the dawn of the life sciences, that the human condition was characterized by "the growth of the natural."[1] The notion of the Anthropocene hadn't arrived at the time (although the signatures were busily brewing throughout the world), but had Arendt followed environmental discussions at the dawn of the twenty-first century, she might have revised her take, alerting us to not only the growth but also the *collapse* of the natural. The escalating extinction of species, in the wake of global heating, rising sea levels, and pollution, is a case in point. The present loss of biodiversity indicates not only the mass disappearance of species but a total collapse of habitat or *Gaia* – of *oikos* (the house) in the classic sense. During the last century, "commons" issues (typically of access and property rights) were usefully fleshed out in environmental discourse. This has drawn attention to growing inequity in natural regimes (fisheries, energy, rights to water, etc.) and the alienation of the poor and the common person. These are issues which need to be addressed by scholars and democratic governments. More broadly, as the Anthropocene takes center stage, it seems time to also reconsider our global house and our place in it.

Indeed, as this book suggests, it is vital to develop a new *Denkstil*, an anthropocenic paradigm. This would necessitate critical thinking and relearning, embracing indigenous voices and qualifying, if not abandoning, the modernist preoccupation with human mastery and total control. An anthropocenic understanding of human beings, several of the authors argue, emphasizes the materiality or elemental

[1] Hannah Arendt, *The Human Condition* (Chicago: University of Chicago Press, 1958), 47.

embodiment of our existence. The Western notion of the autonomous individual bound by the skin fails to capture both our materiality and our relationality. Interestingly, the notion of the individual highlighted in much social theory originally referred to that which is indissoluble (such as the Holy Trinity). Nowadays this is sometimes captured by the notion of the conflated "dividual," a notion that resonates with several indigenous discourses from the Arctic to South America which situate persons and selves in context, socially and materially. After all, our minds and bodies are fleeting and porous, rooted in the (damaged) planet and the Milky Way. Embodiment means that our thoughts and actions are not simply grounded in the ticking brain, but in the engaged body enmeshed in practical tasks. Much of our thinking and enskilment, as a result, is tacit, guided by a contextualized body.

While the notion of the Anthropocene, the *Human* Age, is overly anthropocentric, it should not blind us to the predicament of the more-than-human. Several of the authors of the book under discussion argue that other beings – including other primates, fish, birds, insects, and plants – are sentient, with a life force of their own and deserving space and attention in discussions of the problems posed by current environmental threats. The cliches of the past emphasizing human uniqueness ("we alone have language, think, make tools etc.") have not stood the test of time. The more-than-human speak, reason, and work. How could we add them to our discussions and politics? How might they invite us to their table?

The academic field which I have studied and practiced (anthropology) remains a Western product, despite its rich ethnographies from different parts of the world, its repeated endeavor to avoid Euro-American bias, and its anthropology-at-home. In practice, its theoretical base is that of the North, gazing at the South, reflecting the ethnocentrism and colonial regimes of past centuries. The same can be said about many other humanities, notably philosophy with its strong European roots and its confined intellectual dialogues detached from everyday life, in opposition to the shop floor, "folk" understanding, and indigenous theories. Many anthropologists have argued for a transnational "World Anthropology" to correct the imbalance of their field. Philosophy seems to be on the same course. Collectively, the chapters in this book present some of the seeds needed for genuine World Humanities, cross-disciplinary, cross-cultural, and cross-gendered – and suitable for the turbulent world of the Anthropocene. *Elemental-Embodied Thinking for a New Era* is an important step forward and deserves broad transdisciplinary readership.

Professor Emeritus of Anthropology
University of Iceland,
Iceland

Gísli Pálsson

Bibliography

Arendt, Hannah. 1958. *The Human Condition.* Chicago: University of Chicago Press.

Gísli Pálsson is Professor Emeritus of Anthropology at the University of Iceland. His latest book is *The Last of Its Kind: The Search for the Great Auk and the Discovery of Extinction* (Princeton University Press, 2024).

Contents

Chapter 1
Elemental-Embodied Thinking for a New Era: Introduction to the Volume

Sashinungla, Lenart Škof, and Sigridur Thorgeirsdottir

Abstract A mindful and perceptive partnership between the natural, more-than-human world and human culture is crucial for a connected and ethical way forward for life on our planet. The aim and hope of this edited volume is to contribute to ways for restoring our biosocial relationships and elemental sensibilities—and our forgotten ontologico-environmental, embodied, and material being. The contributors are united by a shared sense of commitment to exploring theories and practices of new sensibilities and methodologies for fresh ways of thinking in the times of Anthropocene crisis. The environmental scholars gathered here come out of diverse traditions of embodied, elemental, and indigenous thinking—and are based in phenomenology, enactivism, mindfulness-oriented phenomenological methodologies, experiential approaches of pragmatism, and indigenous philosophies.

Keywords Elemental-embodied thinking · The experiential turn · The elemental turn · Environmental hope · Partnership with nature · Material being · Indigenous philosophy · Biosocial relationships · Elemental sensibilities · Climate change

A mindful and perceptive partnership between the natural, more-than-human world and human culture is crucial for a connected and ethical way forward for life on our planet. The climate change crisis increasingly demonstrates that we stand at the precipice of a transformation, and we cannot wait any longer to address the

Sashinungla (✉)
Department of Philosophy, Jadavpur University, Kolkata, India

L. Škof
Institute for Philosophical and Religious Studies, Science and Research Centre Koper, Koper, Slovenia
e-mail: lenart.skof@guest.arnes.si

S. Thorgeirsdottir
Department of Philosophy, University of Iceland, Reykjavik, Iceland
e-mail: sigrthor@hi.is

L. Škof et al. (eds.), *Elemental-Embodied Thinking for a New Era*, Sophia Studies in Cross-cultural Philosophy of Traditions and Cultures 42,
https://doi.org/10.1007/978-3-031-42119-8_1

necessary changes required both at the theoretical as well as practical level. We are changing the face of the earth. How humans continue to make that change matters very much in the crisis of the Anthropocene which has been largely produced by human recklessness. The aim and hope of this edited collection is to restore our biosocial relationships and elemental sensibilities—and our forgotten ontologico-environmental, embodied, and material being.

The authors in this book respond to a necessary turn in a way of thinking, because political and "official" solutions to the environmental crisis are not enough. We namely cannot solve a problem with the same tools that created it and a paradigm shift is needed: one related to the elemental and embodied turn addressed in this book. The fact is that, despite green capitalism and UN sustainability goals, the political-economic system of the world steers like the Titanic towards an iceberg.

The editors of this book organized an interdisciplinary symposium on "Surviving the Anthropocene: Towards elemental literacy and inter-disciplinary partnerships." Originally planned to be held at the Institute of Philosophy at the University of Iceland in cooperation with the Institute for Philosophical and Religious Studies at the Science and Research Centre Koper (Slovenia), the Institute for Educational Sciences at the University of Innsbruck (Austria), and University Alma Mater Europaea (Slovenia), the onset of the Covid pandemic in the spring of 2020 thwarted that plan. Consequently, the conference was held a year later, transitioning to an online format.

Participants from diverse parts of the world found themselves locally stranded due to the restrictions imposed by global measures to counter the pandemic. During the online conference, the organizational site at the University of Iceland faced an additional challenge—being shaken by earthquake spells preceding volcanic eruptions that later necessitated the evacation of a town with 4000 inhabitants. The ground beneath was trembling and breaking, serving as a stark reminder of how fragile the ground is on which we stand in so many respects and how transient our localization can be, as evident with the growing number of environmental refugees globally.

The majority of the articles in this book originate from the conference supplemented by some new contributions. The authors assembled here represent diverse traditions, including philosophy, philosophy of religion, anthropology, and indigenous studies. They also represent research teams and networks exploring theories and practices of new sensibilities and methodologies of embodied and elemental ways of thinking in times of pluricrises. The authors are united by a shared sense of being deeply affected by environmentally hazardous events and a commitment to embracing three turns: the indigenous turn, the elemental-embodied turn and the experiential turn.

The Indigenous, the Embodied-Experiential and the Elemental Turn

Firstly, living in the Anthropocene calls for new ways of thinking. The "new" is not something we have to invent from scratch because there have always been traditions of more extended, interrelational ways of thinking within indigenous, spiritual, mystical, and philosophical traditions.[1] Contemporary cogntive sciences are catching up with this fact with their confirmations of how human consciousness is less self-centered and more extended (in embodied and elemental ways) than we have assumed in onesidedly individualistic, disembodied, and disembedded conceptions of man in our philosophical traditions. It is therefore globally important that different traditions be consulted for these insights like is done in the chapters on indigenous traditions in this book. There is a global message here.

Secondly, one may also wonder how patriarchal aspects of main religious traditions are complimented and corrected by the insights into embodied, elemental selves in parts of their traditions that have been marginal or marginalized. This observation has to do with how there is a direct link between a patriarcal sense of divine providence and the dignity of the individual, autonomous subject—which is a good thing. Yet, this fact lacks or underestimates the notion of humans as embodied and elemental beings: beings which are always-already interrelated and interdependent. Environmental ethics and environmental philosophies therefore herald a richer and a more diverse understanding of human beings in a contemporary mindset of disembodied and disembedded understanding of the human. The onesided ways of patriarchal thinking have also been instrumental for colonialization of peoples and the earth, resulting in divisions, wars, repression, and exploitation.

Thirdly, it is not enough to acknowledge how human beings are contextual, earthly, elemental, and embodied. Within ethics and political philosophies these notions have been accounted for within care ethics and political philosophies of natality from Hannah Arendt to Luce Irigaray and deeper notions of life and living (in process philosophies from A.N. Whitehead to Hans Jonas and Eugene Gendlin). As embodied and elemental, we are also capable of sensing and understanding our place in the world—with human and non-human living beings in a different way. Therefore, authors of this book contribute articles about embodied thinking. This is a direct responce to a call for a new way of thinking, for cultivating "anthropocenean sensibilities" and "staying with the trouble."[2] Our desire to stay alive, to thrive (that have been seen as implicit in who we are) can be made more explicit—as a way of thinking that we can tune into and hence better align ourselves with a troubled world. Sensing the misalignment, sensing "the trouble" is where there is a point of departure for acknowledgement and for ways to deal with the trouble.

Finally, elemental thinking is a way of connecting with suppressed levels of living and being. We are touched by the world and we touch it. Human beings necessarily interact with the world and change the enviornment and therefore there is

[1] For the use of the term "indigenous" see Sashinungla's chapter, where she critically approaches the terms such as Indigenous, Aboriginal, Adivasi, First Nations, and Tribal. As opposed to suggesting a new hegemony—in which the diverse indigenous knowledge systems are collapsed into a new universal—in the present book, it is for us important to acknowledge the striking similarities among indigenous cosmologies and ontologies around the world.

[2] The citations are from Nancy Tuana, "From a Lifeboat Ethic to Anthropocenean Sensibilities," *Environmental Philosophy* 17, no. 1 (2020): 101–123 and Donna Haraway, *Staying with the Trouble: Making Kin with the Chthulucene* (Durham: Duke University Press, 2016).

never a pastoral or paradise-type of interaction that we find ourselves in. The forces of the earth are impersonal, with deep time and features that we are powerless against. Manhattan is sinking with its Wall Street model of an economy that is ready to sacrifice the earthly foundations of a species for financial growth that has lost its connection to earthly limitations. We need an economy that stays within the confines of the earth instead of a financial/monetary model that is a mere dance of figures over the abyss. The Earth is asking us to connect. Life wants us to thrive—to sense—because life finds expression in all living.

It is no longer sufficient only to theorize *about* the challenges of climate change. Environmental activism is not raising awareness that results in the systemic and alternative forms of life that are needed. Being the change you want to see is not enough in the face of the enormity of the issues. As environmental scholars, we here take the leap into practices of thinking supported by diverse traditions of embodied, elemental, and experiential thinking—and are based in phenomenology, enactivism, mindfulness-oriented phenomenological methodologies, experiential approaches of pragmatism, and indigenous sensibilities. We try to think and sense our condition from *within* and reflect out of our felt meaning for being one with the environment—by contributing to and being subject to environmental processes and not being separate from them. A new understanding of our place as a species among other species on this planet, and in the cosmos—carried by a self-empathic approach and a relational understanding of an interdependent and interrelational self—may perhaps aid us in coming to our senses in the colloquial sense of finding our bearings within a world that needs healing in so many ways.

Before delving into how the various authors in this book address the challenge of adopting a more embodied, elemental, and experiential perspective, we, the co-editors, would like to cordially thank our Springer editors—first of all Purushottama Bilimoria as "Sophia Studies in Cross-Cultural Philosohpy of Traditions and Cultures" series. Additionally, we want to acknowledge Christopher T. Coughlin, Senior Editor for Philosophy and Religious Studies, for his unwavering support during the planning and preparation phases of this volume. Special thanks are extended to Rahul Sharma for his invaluable assistance throughout the production process. We are also grateful to Vadim Gershteyn for his diligent help with the editing of our essays.

The Elemental

In his "Vegetal Pedagogy" chapter of this book, Michael Marder asks what if the materiality or the matrix of humanity could be viewed as resembling living orchards? With this beautiful metaphor, Marder follows Fyodor Dostoyevsky's "The Earth and Children" diary note from 1876 in which Dostoyevsky states: "I do not draw any differentiation between the earth and children" (Chap. 8).[3] According to Marder, "like trees and grasses, bushes and flowers, children are the produce of the earth in an original take on the Biblical creation narrative, according to which Adam and Eve are made from the earth itself. Pedagogy becomes a matter of cultivation, of unobtrusively channeling children's growth—physical and psychic—in a way similar to the cultivation of orchard trees or fields of cereals" (Chap. 8). What if the task of humanity in an era of the Anthropocene is to relearn to become children of the Earth again, to learn anew these forgotten and ancient teachings of the *materiality* of our existence—through the elemental matrix of our world—in order to be able to secure the future of our planet: the only one that we have. For Marder, a more-than-human pedagogy is required for this decisive task of our era and it is from the elemental perspective of our being that we wish to begin this concluding reflection.

Elemental ways of being are transformative in many ways and we can find ways to awakening of our elemental memory: every breath we take invites us to reopen our bodily perceptions towards this elemental atmosphere in which we are immersed without knowing it. But, not only breath—it is food and its taste, smell of the sensuous world, our bodily proximities towards other human and nonhuman beings, plants and elemental world as such, including stones—that invigorate us in achieving an elementary act of relating to the world as Donata Schoeller beautifully exemplifies in her essay in this book. We *are* the elements, indeed: as eighty-one elements constitute natural matter on Earth, the life of a human being depends on around sixty elements that are found in our body, among them around twenty are essential for our life (cf. Chap. 6). The natural elements thus form a material nexus or elemental matrix on which all life on Earth subsist. We are materially-elementary intermingled with the physical environment and to forget this means also to abandon the elemental ontological grounds on which any human being subsists and to undertake a path of elevating human being above Earth and all creation.[4] This is how we have become despots over living and sensing beings: instead of cultivating an openness and wonder of a child in ourselves, we have rather become modern, old, and cruel. But, even the world is not enough, the systemic forces make us insatiable and wanting more—it is not an ought to have more that guides our crooked mission on this planet for it has muted into a need to destroy to make place for our species.

It is for the purpose of avoiding environmental degradation that the elemental ways were present and safeguarded in all ancient and premodern cultures of the world. The elements of air, water, earth, and fire, and sometimes ether or space as

[3] All references in parentheses are from the respective chapters in this book.

[4] Cf. Merleau-Ponty's thoughts on flesh as an element: "The flesh is not matter, in the sense of corpuscles of being which would add up or continue on one another to form beings. [...] The flesh is not matter, is not mind, is not substance. To designate it, we should need the old term 'element,' in the sense it was used to speak of water, air, earth, and fire, that is, in the sense of a *general thing*, midway between the spatio-temporal individual and the idea, a sort of incarnate principle that brings a style of being wherever there is a fragment of being. The flesh is in this sense an 'element' of Being." (Maurice Merleau-Ponty, *The Visible and the Invisible*, transl. by Alphonso Lingis, Evanston, IL: Northwestern University Press, 1968, 139–40).

the fifth element, constituted the matrix of being on which all life subsists. The liminal spaces between inanimate and animate matter were blurred; rather, the existence was viewed as a flow. This involved the merging of the elements into their ontological habitats, and simultaneously the melting of our alleged essences back into their elemental entanglements. As Guðbjörg Jóhannesdóttir succinctly observes in her insightful essay on glaciers: "me and glacier are for example both made of water, both of us are at the same time solid and liquid forms, we are processes, forces and powers like anything else on the planet" (Chap. 4). In this view, the elemental character of the world of dwelling embraces the mineral strata of nature (stones, metals, minerals…), which are normally understood as inanimate or inorganic nature, and expands towards most complex elemental entanglements. With this new elemental consciousness reactivated through embodied and elemental sensing and understanding, nature reveals as a space permeated by the omnipresent life force of all living things.

Either in their physical (elementary particles as units of interaction or quanta), chemical dispositions (as elements of nature), or in their ancient sense (viewed as forces but also as elemental substances), these elemental webs now enable us to step out of inherited ways of living and expand our sensitivities, perceptions, and thoughts towards what Karen Barad calls an ethics of worldling.[5] This effort can be achieved through re-enchanting of 'lived' materialities, such as stones and forests—as exemplified in Lenart Škof's and Anne Sauka's chapters on pre-Christian European cultures in our collection. It is in this sense, that "the stone refers to essence as slow time that is nevertheless in process and transformation […] [It] represents the potentiality of life within non-life, but also allows understanding of naturecultures beyond the dialectic of civilization and wilderness" (Chap. 3). As further argued by John Caputo, "[O]ur bodies are the issue of solar, galactic, and cosmic stuff, the offspring of the elements. We are all stardust, made of ancient particles."[6] In this view, various lifeforms and their being become incarnated in the *heart of the matter*, with the sacredness of every living being testifying to a newly revealed ontology of matter. This is precisely what is at stake in Emil Månsson's essay on dancing with stars and elements, as he states: "because our bodies are intimately, ontologically linked to elements integrated from our surroundings, which are ultimately derived from an interstellar time before biological life (on Earth, at least); because we depend on the Sun's nuclear furnace and solar wind for our sustenance, and so on […] because our thoughts and hearts go out to them, and our thoughts and hearts *are* our bodies" (Chap. 10).

Finally, to sum up our thoughts on the elemental dimension of our being: as predominantly earthly beings, we are bound to gravity like Ole Sandberg discusses. As an "elemental ontological force of the world," gravity radically shapes and bounds our physical and psychical dispositions and habituations (Chap. 7). Even if we have wished, we still could not have escaped this ontological nexus to the earth in us.

[5] Karen Barad, *Meeting the Universe Halfway: Quantum Physics and the Entanglement of Matter and Meaning* (Durham and London: Duke University Press 2007), 392.

[6] John D. Caputo, *The Insistence of God: A Theology of Perhaps* (Bloomington and Indianapolis: Indiana University Press, 2013), 250.

Gravity is indeed implied in our bodies, as shown by David Morris's *The Sense of Space*.[7] A sense of gravity remains ennatured in astronauts as they find themselves in the situation of weightlessness when they seek orientation, search for the "up" and "down," experience as they were, as it were, "falling" into that vast space. But, there is another possibility of interpretation of this elemental memory: in an archaico-cosmic sense, we might—with Månsson—now play a little bit with his metaphor of dancing with the stars. Is there not another possibility of approaching the primordial sense of the elemental precisely when our bodies are experiencing the weightlessness—as when we are immersed in the medium of water, air, warmth of a light (fire), or earth (such as having parts of our body immersed in sand)? Like astronauts, expanding into the void, being attuned to the echo of primeval silence and vastness of the ancient cosmic spacetime; but perhaps, for the moment at least, experiencing the archaic elemental presence. We are elemental beings and we can say it with Merleau-Ponty in the following way: *the flesh of the world is not matter, is not mind, is not substance...* Indeed, we are the offspring of the elements.

The Embodied-Experiental

The environmental challenges faced by the global population share several common features. One prevalent characteristic among these challenges is the presence of a cognitive dissonance which seems to be a structurally inherent feature in how we experience them. The enormity of the problem of climate change is difficult to grasp and imagine; therefore, individuals feel that their efforts of doing their share remain insignificant in the face of such a massive global problem. It is impossible to achieve an environmentally sustainable life style due to our dependency on technologized systems and processes that are inherently unsustainable. While some progress is made to reduce environmentally-damaging practices and decrease reliance on fossile fuels, the global ecological debt or the unsustainable use of the Earth´s resources, continues to increase. Each year, the Earth-Overshoot Day arrives earlier, such that an ever greater portion of the year is spent living beyond the planet´s means. Under the conditions of the climate crisis, humans are in many respects living in contradictions.

The contradiction that is experientially most significant is that, even though we know the facts about climate change—be it the level of carbon emissions, the increase in extreme weather events and climate related disasters, growing number of climate refugees—many hardly sense climate change directly. The knowledge about it does not yet hit home. There is a delay in effects, similar to overexposure to direct sunlight without the effects on the skin becoming visible or a health hazard until two, three, or four decades later.

[7] David Morris, *The Sense of Space* (Albany: SUNY Press, 2004), 152. For more on this see Sandberg's chapter from this book.

For many, there is nevertheless doom in the air, creating a kind of "pre-traumatic stress disorder" as Sandberg poignantly remarks in his text (Chap. 7). One reaction to this is trying to live well in a dying world, which Månsson discusses in his contribution; this represents the voices of those who are most intensely concerned and already leading lives that prepare for impending social and ecological collapse. Among them are members of the younger generation that heralds the Fridays for future movement—embracing environmentally more friendly styles of consumption and ways of life.

For the majority of people, however, the knowledge of the facts and the sense for the urgency of the matter have not forcefully come together so that the practices of those who are concerned and those on the other side of the spectrum denying any urgency is not so great. Therefore the knowledge factor, i.e., scientific knowledge about the possible and likely affects of climate change does not result in great changes in behavior. The science of climate change is nevertheless one of the greatest polarising issues in public poltical debates, with the deniers who distrust climate and environmental scientists on one side of the spectrum and climate change activists like Greta Thunberg constantly urging people to listen to the scientists on the opposite site of it.

Despite the growth of environmental humanities, environmental ethics, animal ethics, environmental art, and greater knowledge of indigenous philosophies a deeper understanding about the interconnectiton of human- and non-human life has not become widespread. That makes one ask if dominant conceptions of "nature" are still tied up with a mind set that is not conducive to change, and more importantly part of the problem because a concept of nature is tied up in a dualistic and hierarchical distinction of dominating culture (human) against subjugated nature.[8] Even the flurry of films and other products of popular culture that show how a change of mind is required have not been able to turn the tables around. Perhaps the greatest reason is that the overarching financial-political systems are so all pervading and proceed like a Titanic steering directly into an iceberg that it creates a paralysing and resignatory attitude or a sense of desperation.

Given that knowledge and science does not result in the type of profound change needed, climate and environmental scientists and environmental thinkers increasingly call for a different way of thinking and different sensibilities.[9] The collection of essays offered here respond directly to this demand, taking as its point of departure a problem so big that it necessarily leads to cogntive dissonance, paralysis, denial, pessimism, or panic. The articles hence have a common point of departure in these dissonances, beginning where the issue of climate change is sensed as bigger, more complex, paradoxical, and contradictory than our prevalent ways of thinking can easily "hold." The problem can not be conquered with the hierarchical and dualistic ways of thinking that created it. For that reason Treena Balds and Timothy

[8] Timothy Morton, *Ecology without Nature: Rethinking Environmental Aesthetics* (Cambridge, MA: Harvard University Press, 2009).

[9] See Tuana, "From a Lifeboat Ethic to Anthropocenean Sensibilities."

Morton talk about a different way of thinking as a "truthfeel" that "depends on letting oneself be vulnerable, on letting oneself stare calmly at a wall of seeming nonsense." (Chap. 5)

A turn to embodied and experiential thinking is therefore seen as a way of at least felt sensing and realizing the dissonance in common dispositions towards climate change. We not only begin with the "trouble" of climate change, not only thinking *about* it by way of analysing it, but staying with it, trying to gain an understanding of it by way of our felt meaning for the issues at hand like Thorgeirsdottir discusses in her article on how her teaching of Environmental Ethics has incorporated embodied thinking and experiencing. (Chap. 12).

Now, what does this transition from mainly thinking scientifically and scholarly about the trouble we are in to a staying or thinking with the trouble imply? It means taking note of the fact that embodied beings are feeling, sensing beings who are touched by the world and in touch with it. Like Barad notes, "we don't obtain knowledge by standing outside the world; we know because we are of the world."[10] The body is the very link to the earthly environment and discoveries in cognitive sciences of the last decades have shown us the thinking is not only embodied as an intertwining of cognition and affect, thinking and feeling. Embodied thinking is embedded because the ego or the core self is one aspect of a self that also includes an environmental, interrelational, and cosmic selves, and even levels of a vegetal, mineral self that one can sense. As embodied, human beings are life processes that are embedded in all kinds of elemental processes, being touched by different kinds of man-made and non-anthropogenic environments and in touch with them. Developing methodologies of embodied and elemental thinking can issue a greater awareness of how one´s existence influences and is influenced by life processes. Embodied thinking can play a crucial role in realizing how we are going against ourselves as species and other living beings. Connecting in such a way may also counteract a panic mode that is in effect a mode of disconnection. To mobilize transformative powers necessary for significant systemic changes to be made, we can cultivate a foundational sense of connection—an understanding that is embodied, elemental, and experiential in response to the challenges at hand.

The Indigenous

We are at a point in human history when no society can afford to live in isolation. Despite all the *myths*[11] that have been circulated about the 'self-sufficiency' of societies geographically isolated from each other and from the broader ecologies within which we are embedded, the discovery of emerging affinities amongst a group of

[10] Barad, *Meeting the Universe Halfway*, 185.

[11] This is the myth about the divide between man-nature, mind-matter, fact-value, science-faith, man-woman, young-old, and so on.

scholars for whom boundaries (geography and thinking) have ceased to interfere with thought processes is, in itself, a beacon of hope. In the midst of all the plurality, the realization of—and an emphasis to overcome the insufficiency of—the rational-universal/binary-dualist approach is the common thread between the different essays contained in this volume. Affirmation of the uniqueness of human freedom and of the superiority and autonomy of human rationality in modernity is so complete that it translates into extreme differentiation between human beings and other living things—where humans and "nature" are separated by a radical discontinuity.[12] The separation and discontinuity take on a character of ultimacy which affects all our ways of organizing our experiences of ourselves, with others, of the world, and of understanding our predicament. Hence, several of the essays in this volume contemplate reconciliation to overcome the modern fragmentations by examining indigenous philosophies. Indigenous connectional metaphysics[13] unites all beings with nature's activities and forces without fragmentation of experience or knowledge into mutually exclusive realms, like we discuss in this book. Such critical engagement, with the idea of the elements in the indigenous elemental traditions, can re-orient our fundamental understanding of oneness and connectedness with the human other and every creation of Nature. For that, Sauka powerfully instantiates Latvian folk epistemologies as the right ground for thinking about "affirmative environmental ethics" in her contribution in this collection (Chap. 3). The Slovenian Natural Worshippers' insights and teachings about nature model the kinds of narratives necessary to take responsibility for the world around us, as Škof discusses in his text in this volume.

There is desperation and boldness in the contemporary humanistic and scientific community in its attempt to pay attention to the ontological connectedness and the entanglement of everything that exists on Earth. The clarion call is for a different way of being human and relating to the world on different terms. Škof succinctly sums up this indigenous-elemental (and, consequently embodied-experiential) turn in his article in this collection: "life on our planet will only thrive if we are able to understand these sacred old traditions in all their complexity and deep interconnectivity with nature and link them to the findings of modern science" (Chap. 2). It is a myth to think that humans are disconnected from the natural world. To imagine so is artificial, and it is a distortion that minds can be separate from nature. Stressing about the peril of denying ecological relations, trans-species psychologist, Gay Bradshaw has said, "[T]he present planetary state that we wish to heal has been achieved by *denying connection*."[14] By fostering an entire knowledge eco-system based on the predominant genealogies of disembodiment and human detachment

---

[12] Australian environmental philosopher Val Plumwood uses the expression "hyperseparation" to describe such radical dichotomy between humans and nature. See Val Plumwood, *Environmental Culture: The Ecological Crisis of Reason* (London: Routledge, 2002), 123–142.

[13] As we have shown, indigenous creation narratives are a manifestation of a broader metaphysics of ecological connectivity.

[14] Gay A. Bradshaw, "Living out of Our Minds," in *The Rediscovery of the Wild*, edited by Peter H. Kahn Jr. and Patricia H. Hasbach (Cambridge and London: The MIT Press, 2013), 134.

from nature and its elemental connectedness, we have not only endangered our planetary Home but have completely distorted and impaired our children's ability to think, feel, and act, more so in ecologically sustainable ways. While underscoring the role cosmology plays in orienting the lens through which we not only understand the world but also determine our actions, Robin Krimmerer writes about her students' ineptitude to depict any positive imagery between people and the land: "As the land becomes impoverished, so too does the scope of their vision [...] How can we begin to move toward ecological and cultural sustainability if we cannot even imagine what the path feels like? […] These students were not raised on the story of Skywoman."[15] Uncontrolled human ego and greed, backed by a false sense of separation, has led to ethical distancing and legitimised an existence of psychological dissociation.

Cosmology, no matter how distant it may be from our consciousness, dictates the course of our reflections on almost any subject and sets the terms of reference in the ontological, epistemological, and axiological schema of any system of thought. Based fundamentally upon a core understanding of an experiential origin-in-the-sentient-land, multitude of global indigenous worldviews do not close down the ontological possibility for conscious interactions, personhood, and agency throughout all species and across all elemental categories. This embedded ecological thinking is guided by an active attentiveness to the kin relations, as represented by the genealogical bonds (exemplified in creation narratives) that intertwine humans with all creation: bonds from which an attitude of respect arises and guides behaviour towards the environment. Ontology, or values and beliefs about what is possible, what can or cannot exist, and what is allowed is ultimately informed by these cosmological narratives. It is in this context that the significance of Sauka's assertion that "an ontological shift is only possible where it is thought of as part of already existing worldings […] alternate experiential forms […] are to be sought for in the already present, yet alternate imaginaries and worldings that are concealed or less represented […]" is to be seen (Chap. 3).

If what is sought for *and* obligated to is a pedagogy that can account for a more-than-human world to mitigate the precariousness of our current situation, then the indigenous ontologies already include the moon, stone, animals, divine/spirit, trees, and future generations in their sense of "self." Indigenous belonging is ontological (relating to their very being) rather than acquisitive (relating to ownership). So, it is helpful in the way this emulates an approach that encourages people to act sustainably. The most important insight about indigenous environmental thinking is that the tribespeople have a clear notion of responsibility based on the sacred grounding and the connectedness of all beings, and they have a clear understanding of the impact our actions will have on the generations to come. The emphasis on connectedness leads us to deeply integrative questions about our conceptions of 'Nature' and the place and the task of humans in the world—thus inaugurating a fine ground

[15] Robin Krimmerer, *Braiding Sweetgrass* (Minneapolis: Milkweed Editions, 2013), *accessed online:* https://yes-pdf.com/electronic-book/209.

for articulating environmental ethics for the posthuman era. Indigenous elemental stance is also an important position from which to contribute to a larger dialogue and reform of philosophy and science as well as to influence global policy discourse on climate change in the Anthropocene.

Unlike its image and its metanarrative, knowledge (epistemology) is deeply implicated in social hierarchies related to culture, race, and gender, and it serves to reproduce these hierarchies and the underlying ideologies at every level. Therefore, in order for environmental ethics and climate change discourse to be informed/transformed by an indigenous ecological 'sensibility,' it is critical for those occupying dominant/maximal locations in philosophy, religion, and science to attain a capacity to critique their discipline and its practice. There has to be 'uptake' (to use a term from Helen Longino) of criticism, in order to move past the rhetoric of 'respecting difference' to a place of actual pedagogical transformation. Our book contributes towards this course correction.

Given the magnitude of our impending socioecological polycrisis, it is no longer enough to be only *secular* and *rational*. At this point, the kind of sacrality and 'spirituality'[16] that forms the basis of many indigenous knowledge systems across the world has to be drawn upon, for their ethical resources and their emancipatory potential. This has led the authors in this book to turn (back) towards *roots*[17] in search of a comprehensive ethical theory that could account for all aspects of interactions of humans *in* and *of* "nature." This collection of essays is presented in support of and in addition to such explorations and redefinitions. Rather than looking beyond nature to a sphere of ideal(ized) forms of it, we need to collaborate to coexist *with* and *in* nature—as embodied and elemental beings. It is our ethical imperative now to implore our children to look and sense deeper into nature to find the divine spark—a spark that will give rise to a new ecological sensibility which will make possible loving, responsible, and equitable relationships among people and with non-human beings. It is this sensibility that, finally, leads us to try to answer Deborah Rose's poignant question: "How do we learn the attention that would enable us to admit earth 'words' into our lives?"[18]

Bibliography

Barad, Karen. 2007. *Meeting the Universe Halfway: Quantum Physics and the Entanglement of Matter and Meaning*. Durham/London: Duke University Press.

Bradshaw, Gay A. 2013. Living out of Our Minds. In *The Rediscovery of the Wild*, ed. Peter H. Kahn Jr and Patricia H. Hasbach, 119–139. Cambridge/London: The MIT Press.

[16] It is this "conversational" connection and the sense of adaptivity of relationships between humans and their non-human kin, and the natural elements, which grounds tribal (ecological) morality.

[17] Returning to "roots" is meant metaphorically as a turn to embodied-elemental thinking that can reflect a felt understanding of environmental processes.

[18] Deborah B. Rose, "Val Plumwood's Philosophical Animism: Attentive Inter-actions in the Sentient World," *Environmental Humanities* 3 (2013): 105.

Caputo, John D. 2013. *The Insistence of God: A Theology of Perhaps*. Bloomington/Indianapolis: Indiana University Press.

Current climate policies a 'death sentence' for the world, warns Guterres. United Nations. April 20th, 2023. https://news.un.org/en/story/2023/04/1135862.

Haraway, Donna. 2016. *Staying with the Trouble: Making Kin with the Chthulucene*. Durham: Duke University Press.

Krimmerer, Robin. 2013. *Braiding Sweetgrass*. Minneapolis: Milkweed Editions. Accessed online. https://yes-pdf.com/electronic-book/209.

Merleau-Ponty, Maurice. 1968. *The Visible and the Invisible*. Trans. Alphonso Lingis. Evanston: Northwestern University Press.

Morris, David. 2004. *The Sense of Space*. Albany: SUNY Press.

Morton, Timothy. 2009. *Ecology without Nature: Rethinking Environmental Aesthetics*. Cambridge, MA: Harvard University Press.

Plumwood, Val. 2002. *Environmental Culture: The Ecological Crisis of Reason*. London: Routledge.

Rose, Deborah B. 2013. Val Plumwood's Philosophical Animism: Attentive Inter-actions in the Sentient World. *Environmental Humanities* 3: 93–109.

Tuana, Nancy. 2020. From a Lifeboat Ethic to Anthropocenean Sensibilities. *Environmental Philosophy* 17 (1): 101–123.

Chapter 2
Tribal Elemental Ontology

Sashinungla

Abstract In this essay, we explore the tribal elemental ontology through examining such accounts of human-elements-environment exchanges and relations as recorded in their genealogical myths and its associated events and practices. In many indigenous worldviews, humans are conceived as sharing a close elemental communion with other species, animals and the natural world—all forming an important agency of the worldview production. These accounts of life are vastly different from how knowledge is conceived within western worldviews or epistemological understanding. Such metaphysical orientations lead to different types of ontological orientations, specifically with animals and natural elements. The essay looks at the portrayal of the elements within the tribal systems of thought and considers its philosophical and sociological meanings in relation to the broader environment. In particular, it builds upon the role of elements in tribal metaphysics in conceiving the nature of the being, other species, and perspectives on reality or truth claims which in turn presents critical insights for rethinking about our cosmological assumptions and ethical relations with each other and with other life forms on earth.

Keywords Tribal elemental ontology · Natural elements · Elemental tradition · Tribal philosophy · Elemental being · Ontology · Tribal lifeways · Environment · Tribal epistemology · Connectional metaphysics

---

Sashinungla (✉)
Department of Philosophy, Jadavpur University, Kolkata, India

L. Škof et al. (eds.), *Elemental-Embodied Thinking for a New Era*, Sophia Studies in Cross-cultural Philosophy of Traditions and Cultures 42,
https://doi.org/10.1007/978-3-031-42119-8_2

Introduction

In the opening chapter of this volume, we explore a tribal[1] elemental ontology through examining such accounts of human-elements-environment exchanges and relations as recorded in their genealogical myths and its associated events and practices. The examples are drawn primarily from lived-experience as well as from ethnographic accounts of human—elements-environment relations within indigenous contexts through an analysis of the cultural and religious philosophy of the Ao tribe of India's Northeast.

Generally, knowledge is understood within Western worldviews or epistemological understanding. Reason and rationality, upon which this western knowledge[2] is founded, have powerfully characterised the nature of human existence as individual, in society and in the universe. A central tenet of western knowledge is the dualist view of nature and society— hinged on the idea that human is at the center and that nature should be tamed, controlled and subjected to human wills. Thus, a great deal of contemporary environmental discourse is premised on this dualistic assumption of nature-society dichotomy—to protect "nature" from "human." However, many indigenous worldviews tell of a different reality in which humans, conceived of as "spiritual" and "intuitive," share a close elemental partnership with other species and with the natural world. Natural elements, animals, divine, celestial bodies, other species and humans are active participants in mutual life-giving relations teeming with communications and interactions. These metaphysical orientations lead to different types of ontological orientations, specifically with animals and natural elements. This chapter draws upon a range of tribal ontology in order to imagine the human being as a part of the macro-organism, the life system as such, which is a fine ground for socio-ecological thinking. In particular, it builds upon the role of elements in tribal metaphysics in conceiving the nature of the being and the perspec-

[1] Terms such as Aboriginal, Adivasi, First Nations and Tribal are commonly employed to designate indigenous/pre-Christianity traditions globally, with often essentially racist connotations. In India, the term "Tribal" is a British colonial coinage used to represent the "Mongolian" tribes of India's Northeast as binary opposites from rest of India, represented through categories of caste and religion. Indian nationalist historiography adopted this orientalist approach without dismantling the distinction and constituted the Northeast tribes as different from the Indic civilization. See in this regard Sashinungla, "Thinking, looking and acting: beyond East and Southeast to the 'other Asia,'" in Johannes Dragsbaek Schmidt and Shantanu Chakrabarti eds. *The Interface of Domestic and International Factors in India's Foreign Policy* (London-New York: Routledge, 2021a), 253. However, as "Tribal" has entered into common parlance while describing ethnic communities in India's Northeast, it is employed specifically as an analytical category to signify a system of thought–the Elemental tradition, in this essay. As opposed to suggesting a new hegemony, in which the diverse indigenous knowledge systems are collapsed into a new universal, it is important to acknowledge the striking similarities among indigenous cosmologies and ontologies around the world. It is in fact helpful to provide descriptions of the shared components of indigenous philosophy.

[2] The evolution of knowledge traditions now perceived globally in the singular "western knowledge" is too complex to trace all the origins of that knowledge system. Yet, at the heart of the entire process was a conception of knowledge as an intellectual outcome, a body of verifiable abstraction.

tives on reality or truth claims which in turn, present critical insights for rethinking about our ethical relations with each other, and with other life forms on Earth.

An examination of their sensibilities and orientations toward the physical world and especially the elements, will testify to the undying vitality of indigenous traditions in their search for the truth. The ultimate purpose, however, of an inquiry into the idea of the elements in the Ao paradigm is not simply to see the manner in which nature—natural elements—have been conceived within tribal contexts. Hopefully, it will lead this thinking back to renewed understanding of and critical engagement with them. And will re-orient our fundamental human relationships we have with the broader ecologies within which we are embedded. What is of significance here is the spiritual narrative of the tribal cosmogonies (not the literal truth of their content)—the spiritual connectedness and bonds between the people and the whole elemental matrix that it instructs. The edict for how to *live well* on the planet together—between humans and their non-human kin, and the natural elements, characterised by mutual respect, reciprocity and interdependence are lucid in the tribal elemental tradition. Thus, it serves as a template to grasp the meaning of human existence in a network of complex ecological relationships. As Sujata Miri proposes:

> The Adivasi traditions can provide on alternative worldview for contemplating harmony between man and nature. Their cultures were inalienably based on respect for and religious veneration of the natural world. There is in the first place an acceptance of a principle of holistic unity. Secondly there is the belief that all things in the world have an intrinsic life force which explains their interrelatedness. Finally there is the insistence on balance or harmony.[3]

Subsequently this opens up the critical necessity of interrogating the simplistic divides, cosmological assumptions and scholarly norms in science, philosophy and religion. The kind of connectional metaphysics that is operative in the tribal paradigm is distinguished from the monochromatic or one-sided understanding of reality. It goes beyond reason, and the mythos-logos divide. Hence, it constitutes a significant contribution to overcome the insufficiency in the rational approach in understanding reality. In addition, the tribal ontology which underlies the notion that everything is related to everything, presents opportunity to revise the conventional ascetic, monistic and dualistic understanding of spirituality or the ultimate. Therefore, the tribal elemental ontology offers a distinctive vision statement for the new era, due to its altogether radical approach.

As a response to earth (socio-ecological) crisis, indigenous and ecological thinkers have called for rejection of ways of thinking and being that lead to separation, domination and destruction. In this direction, we reiterate that indigenous lifeways around the world instructs as well as forms the basis for a renewed understanding of reality—for the personal life of human persons and the collective existence, especially in the context of the contemporary challenge of pluralism. The kind of holistic

[3] Sujata Miri, *Ethics and Environment: Theory and the Adi Khasi Practice* (Guwahat: Delhi, Spectrum Publications, 2001), 87.

sensibility rooted in the tribal elemental ontology may well be the catalyst to seek a new vision, not only in the wholly other (transcendent) but also with the tangible and corporeal other—within the cadences of the human and the cosmos.

Truth and Tribal "Lifeways"

Metaphysics,[4] understood as the first set of principles, or, the foundational categories of systems of thought that cultures or peoples maintain in order to comprehend the world is a powerful concept with which to critically assess a philosophical system's fundamental assumptions and normative frameworks, especially the nature/characterization of reality. If metaphysics is understood as first principles that shape worldviews, then how would a culture's perception and value of human beings' existence as individual, in society and in the universe shape its worldview? Here, in order to situate the study in the context of worldviews that offer fragmented perspectives of world/reality, it is important to examine, at least briefly, the nature of the first principles; as conceived in the western knowledge systems[5] and in the tribal knowledge systems.

Regarding being, or ontology, western views emphasize anthropocentric construction. Philosophers such as Socrates, Plato, and Aristotle, who are regarded as the founders of humanistic tradition of knowledge, promoted the notion that humanity is on a higher plane than other "beings" such as animals and plants because of humans' capacity for language and reason. Whereas, tribal ontological relations include not only people but also animals and whole of *creation*—understood as the

[4] Since the time of the Greeks, both worldview and metaphysics (or philosophy) have implied a "vision of totality." Because of this similarity, worldview and metaphysics are often used interchangeably. But, as the concept of worldview in cultural anthropology has been developed in order to address culture under the gaze of the anthropologist in a field mostly concerned with "observing" the indigenous peoples—the "primitive" and exotic cultures, it does not quite capture the foundational issues the notion of metaphysics does. Therefore, it is important to employ these concepts as analytically distinct categories. Also, the concept of metaphysics captures the primary subject of the present essay's analysis most accurately. In fact, it is useful to understand metaphysics as first principles, which in turn produce/shape (particular) worldviews.

[5] We do not offer a critique on the western metaphysics from an indigenous metaphysical stance, as the literature is already replete. Nor is this a systematic, comparative account of the two metaphysical systems but only points out how some key elements of metaphysics are conceived differently in the two systems, which is pertinent to the essay's analysis. For a powerful critique of western metaphysics from an indigenous (in this case, native American Indian) position, see Vine Deloria JR, *The Metaphysics of Modern Existence* (San Francisco: Harper & Row, 1979, 2012); Vine Deloria & Daniel R. Wildcat, *Power and place: Indian education in America* (Golden" Fulcrum Resources, 2001), to cite a few. For more discussions on indigenous metaphysics from diverse contexts and perspectives, see Linda Tuhiwai. Smith, *Decolonizing methodologies: Research and indigenous peoples*, (London & New York: Zed Books, 2012, 2nd ed.); Anne Waters, *American Indian Thought* (Oxford: Blackwell publishing Ltd., 2004); James Maffie, *Aztec Philosophy: Understanding a World in Motion* (Boulder: University Press of Colorado, 2014); Stephen Muecke, *Ancient and Modern: Time, Culture and Indigenous Philosophy* (Sydney: UNSW Press, 2004).

whole elemental matrix. Every member/component, be it human, animals, divine, plants and physical elements, is entitled to a place in the larger scheme of things—i.e., for the upkeep of the cosmic order and the continuity of all "life." Humans are a key part of the world, as custodians of stories and ceremonies, but it is not centred on them. Philosopher Mrinal Miri's observation here is salient: "The world of the tribesman is seamlessly continuous between the inanimate, the animate and the human; she/he is ensconced in the contingencies of time and space as anything else in the world… The episteme of the tribal vision is similarly continuous between the natural, the moral and the spiritual."[6]

In many tribal traditions, time (a key element of metaphysics) is generally perceived as a cyclical manifestation, linked with the seasonal cycles, ceremonies and practices connected to the "place"–not to the exclusion of linear time. Western views, on the other hand, emphasize linear notions of time–concerned with chronological recounting of historical events or maintaining exact dates. Commenting about the different cultural attitude concerning temporality between the western intellectual tradition and the tribal paradigm, David Scott observes:

> Tribal spirituality and values, social and political structures and even ethics are rooted, not in a temporal historicity, but in spatiality. This is perhaps the most dramatic (though largely unnoticed) cultural difference between the experience and thought of indigenous peoples and the western intellectual tradition… Tribal people have a temporal awareness, but it is subordinate to their sense of place.[7]

In the tribal framework, the concepts are not necessarily word-based, but rather place-based and body-based, and that a place and nature is not simply an object, but it is shot through with life-giving relations—is alive, imbued with "spirit," and a central agent in the worldview production. This notion is vastly different from that of the western metaphysics, where place or geography is an inanimate object, devoid of agency, at the disposal of human intervention— another object simply to be administered or exploited.[8]

Further, western conception of reality with its origin in the Christian monotheistic structures, value a singular or universal truth. This metaphysical foundation produces a peculiar type of epistemology and ontology—with totalitarian principles, operating through such discourses of science, religion, philosophy and politics.[9]

---

6 Mrinal Miri, "The Spiritual and the Moral", in *The Oxford Anthology of Writings from North-East India: Poetry and Essays*, ed. Tilottoma Misra (Oxford: Oxford University Press, 2011), 116.

7 David C. Scott, "A mirror to M.M. Thomas's Perception on Inter-Religious Studies," *The SATHRI Journal*, no. 2 (1993): 87–88.

8 John Locke provides a concrete example of the exploitative nature of western metaphysics in his chapter on property: "God, who hath given the world to men in common, hath also given them reason to make use of it to the best advantage of life, and convenience. The earth, and all that is therein, is given to men for the support and comfort of their being." See John Locke, *Second Treatise of Government,* edited by C.B. Macpherson (Indianapolis: Hackett publishing company, 1980, reprint), 18.

9 Western metaphysical structures not only promote universal narratives, but also impose themselves universally. Even liberalism has the stamp of Christianity's dualism that has never disappeared in the secular attempt to rewrite natural right without God.

This approach of absolute and universal truth claim, central to western epistemology, is vastly different from that of the tribal foundational way of coming to know and understanding the nature of life and our relationships in the world. For instance, the Ao elemental tradition do not conceive of "beings" as the fundamental principle of reality but rather the fundamental principle of reality consists in creation—which includes the whole socio-ecological complex. This kind of conception is different from Aristotle's characterization of what is real (eternally true). Aristotle understands science as a body of knowledge that seeks the eternally true, whereas, the tribespeople understand the character of reality, as it is given; is not eternal. In tribal epistemology, there is no such absolute and "objective truth" or justification, at least, in the sense that no such objectivity is more valuable than the "truth" of the interpersonal-elemental relationship or cohesion and continuity.

Here it is important to underscore how the notions of time, place, reality and being have come to shape key ideas such as dualism, universalism, humanism, anthropocentrism and other related concepts that inform the cultural orientation of the West. These concepts together, create a complex thought system—a system inherently dualist and objectifying in nature. Similarly, western metaphysics promotes a framework that locates the ontology of human beings as oppositional to that of nature—human-nature binary. Such positioning enables the domination and dehumanization of those falling outside the boundaries of western knowledge paradigm, and strips any ontological agency that other entities may have.

The totalizing principles of western metaphysics with its emphasis on universal assumptions, linearity of thought, and the colonizing nature has tremendous implications upon whose worldviews maintain alternative narratives, traditions, and knowledge claims. Different conception of the first principles or the metaphysical orientations not only produces different worldviews but also informs what is considered as standard science or what counts as "true" and "false" knowledge. The rise of Platonic reason and the resultant unstable power relations between various "regimes of truth" has rendered tribal knowledge system to "primitivism"—with the scholarly employment of "folk" narratives or "myth" with other derogatory terms that imply backwardness and the general derision of native lifeways, indicating "questionable accuracy." Myth stands opposed to truth when it is considered as a part of a fiction-fact binary. As drawn above, the value complex that underpins this division into truth and fiction has a long and convoluted history, and is based upon a combination of religious, philosophical, and scientific thought that has been fostered for over two millennia.

Such questions, and the legitimacy of narrative as "fact" or as relevant knowledge, can be rendered very differently within tribal, as compared with "secular" western contexts. In tribal framework, the kind of reasoning employed takes into account all aspects of experiences and interactions of humans *in* and *of* "nature." Unlike the kind of categorization available in Cartesian dualism and scientism, for which subjects are the sole agents and objects are bereft of all agency; tribal narratives involve shifting concepts of personhood and identity as applicable to the "characters"—including non-humans, and a markedly different distinction between fictional and non-fictional. This does not mean that the tribal narratives do not recognize fiction but that it is not employed in the same manner as within western categories of writing and thought. In traditional Ao culture and religion, for example; there is a different mapping of the texts, in which the understanding of the texts is not necessarily a written document. Subsequently, the understanding of the "world" here is much broader. Their synaestheticism is a complex mode of sensory perception, unlike the exclusive preoccupation of the visual sense with the text. An important feature of the Ao culture is oral poetics which highlight their performative aspects.[10] There is certain fluidity and the pluriformity in their text, unlike in the rigidity and the uniformity of the written traditions.

This is neither to suggest that indigenous groups or people are especially more intuitive than others nor is it an attempt to elevate a mythic "primitive nobility," which leads to romantic particularism and tribal isolationism. Rather, it is about pedagogical accountability.[11] The critical narration of a tradition—which is both marginal and oral, without reduction or, through appropriation, is important. Marginal, because tribes exist in the margins of dominant groups and their intellectual frame and worldview; influenced by dominant philosophies and often unable to appreciate and accept what is meaningful in their worldview. Our preoccupation with Platonic and Cartesian questions about knowledge, truth and being should not make us overlook the fact that the tribal ways of "knowing" are distinct from the western system of knowledge who give primacy to reason and objectivity. It is critical to recognize this conceptual difference. Failure to display an equal consideration of different ontology to the tribal metaphysics results in a superimposition of dominant ontology onto a tribal metaphysical system—and is unable to appreciate the immense transformative reality of tribal systems of knowledge. Tribal "lifeways" have to be studied on its own terms, not through the prism glass of "dominant modes"[12] of thinking or categories. Because if the metaphysical difference in worldview is not taken into account, it simply perpetuates the cosmological consensus of the dominant culture and is left unquestioned in areas where it is hegemonic.

There are attempts to bring cultural, religious, philosophical and science's sensitivity to the Earth's (human-ecological) realities in the Anthropocene. However, these efforts remain superficial because it does not unearth the ontological reasons responsible for this crisis that has created an entire knowledge system that divides human from human and human from nature. Faced with urgency, some postmodern

[10] Its performative nature encompasses such elements as sound and music, dance and ritual, oral narratives and dramas, basketry and woodcrafts, textiles and artworks, body art and ornaments.

[11] Dominant frameworks categorize tribal thoughts as myth, folktale or fable, or as genres that are studied within the areas of folklore, cultural & religious studies, literature and anthropology. Hence, tribal lifeworld—its cultural and religious philosophy is not systematically studied as a distinct system of thought/philosophical tradition, in academia. For instance, none of the philosophy departments of all major universities in India include tribal philosophy as part of their syllabus. Therefore, we propose that our engagement with indigenous systems of thought must go beyond the present stereotypical focus about how tribal ecological knowledge be applied to tackle specific contemporary socio-ecological problems.

[12] Tribal religion should be studied on its own terms, not in terms of Abrahamic-Dharmic categories. Its meanings should be brought to bear just as it was/is experienced by people who practice it.

philosophers, points to tribal lifeway which as they see, is a narrative grounded in geography—where our location is understood in the elaboration of relation in a non-essentializing narrative, achieved through grounding in the geography of our lives. Jim Cheney for instance, argued that a principle of epistemological method may be conceived from the—"non-literalist understanding of the experience of the indigenous lives; to interpret the unfathomable depth of stories and experience and everything that human experience has value and instruct us in some aspect of life."[13] The reasoning is that, if value is implicit in our descriptions of the world and our place in it, then the narratives we construct will embody value and orient us.

The idea of learning from animals, or communicating with animals and natural elements, seeing spirited aspects within nature or affording personhood and wisdom to non-human beings in general, breaks a number of conventions that are well established to the point of appearing normative and can present enormous cosmological instability for those trained in western traditions. Humans (modern minds) are encouraged to speak about animals as part of a category called "nature." The natural sciences have no room for concepts of spirit in the world, and Christianity does not have a cosmological precedent for allowing spirits (or souls) in "nature," limiting the ownership of "souls" to human beings. From this conceptual framework, animate or sentient aspects of nature, including natural elements and the concept of other creatures as speaking, acting or as spiritually endowed beings, remained possible only in the realms of myth and fiction.

Ao Elemental Ontology

In the tribal paradigm, subjectivity in the form of sentience and agency is not uniquely a human prerogative but is located throughout all species and across all elemental category. The experience of "language," as something that is shared with all beings or everything that is; not only amongst humans, is a very common feature across indigenous elemental traditions worldwide. Many tribal traditions in India are oriented to a lifeway in which everything—be it animals, celestial bodies, water, fire, earth or the trees and landscape have the possibility of intimate knowing relationships in interspecies/cross-elemental consubstantial kindredships. In the tribal system of thought, the way of being in the world is about recognizing ecological sentient—paying attention to, "speaking" with, and understanding animals, objects and entities of the natural world as part of *living well*. Tribal views perceive natural elements as "alive", imbued with soul or spirit—and all together forming an

[13] Jim Cheney, "Truth, Knowledge and the Wild World," *Ethics and the Environment*, vol. 10, no. 2 (2005): 105. Also C. J. Preston's description of "Contextual Epistemology" is an excellent work to look at in this regard. See "Epistemology and Environmental Philosophy: The Epistemic Significance of Place," *Ethics and the Environment*, vol. 10, no. 2 (2005): 1–4.

important agency of tribal worldview production. An Ao origin mythology illustrates the character of the tribal epistemo-ontology[14] very well. It goes like this:[15]

> Having emerged from the six stones,[16]
> In the very beginning, no knowledge of water humans had
> They drank the juices from stem and trunk of plants when thirsty
> Imsenpirong and Senpirong, two brothers see one day, Bulbul bird shaking off from its feathers a clear liquid
> Seeing the two brothers, the bird started singing, "Imsenpirong and Senpirong look in the creek under the rock"
> After hearing the song, the two brothers looked under the rock and found water
> The whole village rejoiced for the discovery of water.

The possibility of human mind immersing in intimate knowing communion with non-human animals and with any part of physical nature is powerfully described in the tribal genealogical narratives. These narratives recognize difference but the difference does not amount to their being devoid of consciousness and intentionality, as tribal conceptual framework is not limited by the dualistic structure of otherness and negation. Ao ontology, thus expands the concept of "nature" including in it the entire ecosphere. In the tribal paradigm, the universe is "alive"—a hugely dynamic entity to be entered into a dialogue with, to respect and establish a reciprocal relationship with. In that sense, nature is not beyond moral categories. Their legends characterize animals and other natural elements and forces as beings capable of inspiring respect as also disfavour, yet, recognise that each is a different species, though members of the same biotic community.[17]

Ao philosophical lyrics and oral poetics of the pre-Christianity period, speak of trees, sun, moon and mountains as beings who felt, perceived, deliberated and responded voluntarily. It is a common feature among many tribal communities in India's Northeast to have cradle songs for little children that speak of the great respect they have for natural forces. These songs describe natural elements as examples of stability and dignity to look up to. For instance, the cradle songs mothers

---

[14] Shifting between epistemology and ontology without even 'recognizing' the difference, since conceptual necessity equals natural necessity—in the sense, epistemology cannot be investigated without regard to what there is.

[15] The translation is our own.

[16] Reference to the Ao metaphysical explanation of origin, the fundamental principle of order. I elaborate more on this in the following paragraphs.

[17] An Ao origin narrative illustrates this point well. It goes like this: "Once upon a time men did not know what rice was. One day a rat said to a man: "I will give you a present if you promise me a decent funeral when I am dead." The man agreed to this and the rat brought him the present of rice and showed him how good it was to eat. A little later the rat decided to test the man's good faith. So he lay down on the bank of a stream and feigned death. The man came along with his son, who saw the rat and pointed it out to his father. Instead of fulfilling his promise the man said: "what a dead rat?" Poke a stick under it and push it into the stream." The rat, who was not really dead, of course, jumped up in anger and said: "In return for your faithlessness I shall always eat your rice first and leave my droppings in it." Then the rat ran away and jumped into the river." See J.P. Mills, *The Ao-Nagas* (London: Oxford University Press, 1973), 312–313.

sing to little children express the dialogic connection (not their indifference to us), the Ao tradition envision with natural elements. An Ao cradle song goes like this:[18]

> Mother: O moon come down to play together
> Moon: can't come leaving alone my mother!
> Mother: Come down, leave your mother with your elder sibling and come
> Moon: elder sibling gone to fetch water, hasn't returned home!
> Mother: Come down, leave your mother with your younger sibling and come
> Moon: younger sibling gone to gather firewood, hasn't come back home!

This non-binary dualist[19] understanding is in marked difference from the classic Platonic binary dualist ontology of Being (and non-being) or from the Cartesian paradigm of radical separation between man, who had a soul, and the rest of creation, understood as inert material with no spiritual dimension. Duality can be highlighted by the deep belittling of the animals and nature that happened in the first centuries of this era.

Tribal paradigm characterization of nature—natural elements in a broader canvass suggests an understanding of the nature not merely as passive recipients of human actions but as active participants in the whole web of relationships. For instance, one Ao creation story tells of a time when fed up with the continuous griping of humans about sun's heat, the sun refused to come up plunging the earth into darkness and disarray until the rooster was sent to persuade the sun to return. The story provides profound insight about the place of humans alongside, not above, other beings in the cosmic order—all as constituents of respect and rights. Just as humans can alter the world, the world can act on and alter us. In this respect, the universe is personal—that is, having personality and particularity. This kind of interrelatedness and the totality of life to the whole creation informs a particular construct of *being*—one not separated from land/place. The self of a person is not merely an individual but a socio-elemental being, present and active in all parts of the totality. Here the being is conceived as neither substantive (in the Aristotelian sense) nor relational (in the sense that the world comes into being (only) as my consciousness opens up to it). The world has a vitality of its own of which humans are an integral component, but not centred on them—world exists absolutely and beyond human intervention.

This elemental orientation of the Ao ontology is most clearly found in their cosmological myths—which relates the character of the cosmos and the origin of human beings. For instance, the Aos have an interesting myth connected with their ancestral origin in an ancient place called Chungliyimti. According to the myth, the Aos "originated" out of stones at Chungliyimti, tracing all the six major clans as

[18] The translation is our own.

[19] For an analysis of the non-binary dual nature of reality in Ao paradigm see Sashinungla, "God-Self-World Continuum in Tribal Religion", *Poligrafi*, 103/104, no. 26 (2021b): 93–105.

having "emerged" from six stones, called *Lung trok*[20]—symbolising six chiefs or ancestors. What is most significant about the origin myths for philosophical purposes is that it makes very clear that what happens to exist now is an expression of a specific configuration of the divine—each piece is a special configuration in a cosmic order. Ao cosmogonies,[21] by "relating" them to nature —natural elements, thereby establish a connection to the land in the very nature of their being. Land/place becomes the primary referent for all formulations of meaning and value—in essence, how peoples come to know.[22] Hence, genealogical affiliations are not only crucial in understanding of the universe, but also of the nature-society relations in ontological and epistemological terms. There are moral imperatives between individuals (and communities) and the natural world because of their ontological connectedness–that is, a connection inherent in the nature of their being. These genealogical connections are ethical connections—expressed as *sobaliba*[23] in the Ao language. *Sobaliba* embodies the fundamental Ao ethic of reciprocity between people, and between people and land.

To bring out the framework of the Ao elemental ontology further, we expand on the idea of Earth as it is characterized within tribal system of thought and look at its philosophical and sociological meanings in relation to the broader environment. Whether the earth (one of the four "roots" or *rhizomata* as Empedocles called it) is taken as ground or home–our planetary Earth, it is conceived and experienced in many different ways: as soil, as place or land of persons and communities, and as underworld or realm of death. As a classical element in ancient Greek philosophy and science it was commonly associated with qualities of matter and the terrestrial world. In many cultures, the Earth has been personified widely as the Mother Earth

---

[20] A cosmological explanation of existence within land. Its extended phrase *Lung trok ko voker* refers to the mythical age of origin, the primordial time of origin, the time of ancestors which gives meaning and significance to all that follows. Whilst referring to a past, it indicates a Creation Time that is also ongoing with people, place and the ancestors connecting in a co-creative non-linear present. This notion of a people and their ancestors converging or collapsing of space-time continuum in the co-creative present is also a fundamental characteristic of Maori philosophy. See Georgina Tuari Stewart, *Maori Philosophy: Indigenous Thinking from Aotearoa* (London: Bloomsbury, 2021).

[21] Creation narratives also relate humans to animals and other creatures, affirming the belief that they are the progeny of the same ancestors and therefore, are the same being: they are their kin, and some are their totems. The tribal totemic system demarcates a world of "difference," and at the same time weaves patterns that connect particular human groups with particular non-human species and elements, such as an animal, plant or spirit-being.

[22] Writing on Aboriginal Australian cosmology and spiritual practices, Hume calls the result of creation as "geosophy": a sentient land containing wisdom from the past. See Lynne Hume, *Ancestral power: The dreaming, consciousness, and Aboriginal Australians* (Carlton South, Victoria: Melbourne University Press, 2002), 25. Creation stories in Australian Dreamtime cosmological thinking bears similarities to Ao cosmological thinking through *Lung trok*.

[23] *Sobaliba* is the core of Ao people's sustaining principle. It comprises of protocols, both ritual and practical. Such protocols secure the continuance of community life in socio-ecologically sustainable ways. Considered together with *Lung trok*, it is the highest reference for direction—an ethic about how to be in the world.

or as a female deity. In Ao mythology, Earth symbolises Lijaba, the Supreme Being. It also symbolises one's ancestors.[24] Consistent with their non-binary dualist ontology, Lijaba is identified as male but has attributes described as "feminine." Lijaba is literally translated as, "earth" (*Li*) and "indwells" (*jaba*). It means "the one who indwells the earth." The word *jaba* also means "real" or "existence'; therefore, Lijaba could be also translated as, "the real earth" or "embodied being." Thus, the Earth as Lijaba is existence for the Aos. While the same being is addressed by different functional names, they are all addressed as Lijaba, the one Supreme Being, in relation to the creation of the earth and everything therein.[25] Since the earth is not only the dwelling place of Lijaba, but also the place of their origin, it forms a sacred space of communion with the Supreme Being and their ancestors.

Further suggesting that the *naturalism* of the tribal religion—its space oriented theology does not characterize the nature of the Supreme Being as totalitarian. The Supreme Being is not seen as directing the process of the cosmos as an unmoveable mover, who is absolute up there. Though Lijaba, the Supreme Being is understood as creator and sustainer of all things, it is also perceived, though distinct, as part of the total eco-system. In the tribal religion, God is not the central point of reference in understanding all reality, nor is the world centred on humans. Rather, together with God, humans, animals, nature and the whole elemental complex constitute a cosmic web of relationships in which the truth is to be lived with all others. This account of tribal naturalism is different from the western naturalist's position[26] as the latter does not require any specific sort of religious commitment. In the tribal account, it never loses its connection with divinity.

Ao elemental paradigm conceived the land as both a space and a place. Land is a space because the tribal philosophy emphasis the awareness of space as a basic category of life's orientation. It means that everything is integrally related to one another and therefore, they must treat each other with justice, required and demanded of each other. Space represents the nature of reciprocity between humans, between human and animals and the rest of creation as grounded in common relation to the land. Whereas place connotes "home" and is bound up with a sense of belonging and a narrative which conveys an identity and provides a basis for participating in

---

[24] Beside the Aos, in the creation myths of many tribal communities of India, they consider themselves to be "children" of the earth. For instance, according to Tangkhul tribe's legend, humans originated from the bowel of the earth called *Miwurlung*.

[25] The prayers offered during the Lijaba *mong*/Lichaba *za*, a special ritual observed by the village elder placating Lijaba and seeking forgiveness for any lapses in conduct of the Tsungrem mong festival expresses the great respect the Ao tradition has for natural elements or forces. Tsungrem mong is a 3 days sobriety ceremony addressed to Lijaba for good weather and bountiful harvest. See in this regard a remarkable work about Ao indigenous religion and culture by Panger Imchen, *Ancient Ao Naga Religion and Culture* (New Delhi: Har-Anand Publications, 1993), 55–57.

[26] Naturalist position consider nature as an ideal or an object of inspiration for humans to follow—to find beauty, source of wisdom and ethical instructions. Modern environmentalism has been based on this hope of naturalism. However, it is important to notice that this form of naturalism is premised on the nature-society dichotomies. But such understandings are incoherent in a world made up of dynamics of ecologies that have mixed with human across centuries.

history. From this elemental stance, every river, every mountain, and every peak has a legend and every terrain has a tale to tell. Their traditional form of greeting "*Na koyimmer*"? (literally translated as, "From which village are you"?), when two Aos meet for the first time, is suggestive of their definitive belief that self is a bio-geo-physical being. For in their thinking, (place) village is not simply a geographical location, but the site of his existence and the milieu which shape his identity as a (rational) person—thus, cannot be disconnected from it and the greater environment, and even more obviously from each other. Ultimately, in the traditional Ao ways of being; a person's identity as well as his spirituality manifests through his relatedness with the "living" land. Further, the village as a place of inhabitation, or the act of naming the village also represents a spatial feature. For instance, a village such as *Yaongyimsen* recalls the memory of long-departed *Konyak* settlers with its new connotation as "new village of the Yong people." The Ao way of naming reveals a synchronism between the present and the past.

In the Ao epistemo-ontology, there is no simplistic distinction between living and non-living but only different forms of life. Everything, land, moon, animals, trees, rocks have *temela* (spirit/soul). The world is saturated with life, and death is not conceived of as "separation" but as passage to *Dipuyim* (translated as "land of departed"). There is only a transition of the present life to the next life in the *Dipuyim*. Therefore, their relationship with the land (Earth) does not cease in "death" but continues—eternally bound in the sacred web of connections. Unlike in most universal-rational cultures, the Ao ways of being do not conform to an animate concept in relation with a presumed inanimate opposite. Their treatment of *sungkah* ("dead" trees that are left either standing or lying down to decompose naturally) demonstrate this point very well. A tree may be dried and shorn of leaves but it is not considered "lifeless" as it still supports life and forms an important part for the flourishing of the total ecosystem. Besides the birds for whom it is home and a place to nest, it also allows other life forms such as fungi, insects and worms that form the web of life called (home) nature. As we have argued elsewhere, unlike Aristotelian radical binarism ("true" or "false"), tribal conceptual frameworks express ontological nebulosity: it does not exclude the in-between ("neither" or "sort of") proposition; and recognizes that more than two values can be accounted for, without leading to radical exclusion or contradiction.[27]

As kinship system extends into the land and the land has *temela* in Ao ways of being, land is not an unattached entity of the universe. Interestingly, Ao lexicon does not have a term for something called "untouched nature," at least in the sense modernist characterize the natural world in terms of "wilderness" or "virgin forest."[28] There is no such uninhabited nature, empty or vacant land merely for territorializing and management. In the Ao elemental tradition, the land with nature as a whole is home for all forms of "life." From this perspective, land (the whole Earth) is the source of life (site of origin), dwelling place of their deities and venue of worship,

---

[27] Sashinungla, 2021b, 94.

[28] In fact, ecosystems are result of dynamic interactions between societies and nature.

and the abode of their ancestors—the Earth is Lijaba.[29] Land is the material element that unites Lijaba, humans, their ancestors as well as the animals and the other beings as one community. Since no dimensions of life is divorced from ecology in the Ao conceptual framework, nature is not something we *relate*[30] to, but being in nature and being part of it. Appearing in an Ao philosophical lyric, the following verse convey this point very well[31]:

> Oh! In the God's dwelling place *Aliyenden*[32]
> Where sun and moon are revered, and
> Fig tree bear fruits in abundance
> The joyful dancing of youngsters continues in celebration always.

This conception of a continuance between humans and nature in the Ao philosophy is thoroughly religious (reverence) as much as it is for pragmatic (utilitarian) principles. The premise is clear: to sustain the underlying pattern of well-being (body and soul), resilience and adoptability that maintain the planet in a condition where all life as a whole can flourish.

This awareness of connectedness between the human spirit and the natural life force in tribal philosophy, constitutes the fundamental understanding about what it is that makes us human. Therefore, tribal creation narratives are a manifestation of a broader metaphysics of ecological connectivity. *Sobaliba*[33]—the Ao edict of life, points to collective responsibility for land and others, and a deep gratitude and humility of what it means to be human. When combined together with *lung trok*, *sobaliba* is both an action guide to living righteously/living well and a guide to understanding reality, in relation to Nature (both social and physical) as the basis for all meaning.[34] There is "life" in every element of nature and all life is sacred. Therefore, violating them becomes the violation of the very conditions which makes humans human! Humans well-being and self-respect cannot be understood disconnected from the well-being and dignity of the whole creation. And, to "separate" human from nature and the Divine is to lose the meaning of life or for Plato, the very *excellence*[35] of human life. Thus, in the Ao philosophy, good/truth is not only an abstract concept, but a way of living that comes out of meaningful, reciprocal relationships with the entire environmental-elemental complex that makes life possible.

---

[29] The earth is the ethical imperative. The traditional Ao practically testify on the very earth they stand. In fact, a typical *Mezungtep* (translated as swearing) invokes: "by the very ground under my feet as my witness." A valuable work about Ao customary law see Tajen Ao, *Ao Naga Customary Laws* (Mokokchung: Aowati Imchen, 1980).

[30] Because we relate to something outside of us. But we are in nature.

[31] The translation is our own.

[32] Translated as the village assembly place (like the Greek's "agora").

[33] An important work detailing about the Ao sobaliba see N. Talitemjen Jamir, *Asen Kin Sobalibaren* (Kohima: N. Talitemjen Jamir, 1986).

[34] More discussion on this, see Sashinungla "Ao Worldview," in Sujata Miri ed. *Rationality and Tribal Thought* (New Delhi: Mittal Publication, 2004), 175–180.

[35] Almost all Greek moral theories consider the nature only as a means to realizing the excellence of human life, which is the highest purpose in the hierarchical, teleological order of nature.

This is quite different from how Plato idealised abstraction and a life free of material restraints and distractions.[36] The principle that *everything is connected* suggests that truth is not something figured out abstractly in the mind alone; it emerges under a certain cosmological configuration that can only be fully understood in a connectional and ethical context.

What is emerging from the discussions is that, in the tribal conceptual framework, reality is seen as a web of intimate ontological relationships that cuts across difference—human, animals, other species and all elements of nature. This earth-oriented approach to reality is based on the fundamental premise that human is a part of the physical nature having an inalienable and mutual interaction between them, thereby, relationships characterised by both mutual respect and mutual use. Thus, tribal elemental ontology unites human with nature's activities and species in a bond of a mutual cycle of life-giving continuity and does not allow the fragmentation of experience into mutually exclusive realms. The basic principle of reality (Lijaba), characterised by a kind of non-binary duality: human and nature, male and female, material and spiritual, transcendence and immanence, and so on; exists as a connecting relation. From this, a relationship of reciprocity is established across categories of meaning, never losing sight of an ultimate wholeness. It is this very condition that makes possible communication and love for the "Other" without undermining the reality of the other.

In Conclusion

The "conversational" connectedness and the sense of adaptivity of relationships across difference/species, which is at the heart of tribal elemental ontology; appreciates the mythos without reducing it to any particular form of epistemological medium created by the logos. Tribal ontological nebulosity embraces complexity and thus, accepts the probability that there can be a variety of different standards and reasoning for determining truth, thereby, opens up the possibility that one can have (or not yet have) knowledge and truths that are not familiar to or possessed by others. Thus, knowing the limits of one's own knowledge and being willing to share and listen to what's shared are integral to supporting such a principle.[37] In that sense, tribal philosophical traditions have a concept of truth that is both universally applicable but also eminently revisable. Therefore, the emphasis in the tribal accounts of how "knowing"[38] happens when it is not limited by mental considerations or assumptions regarding the nature of the universe is salient. Unlike the old scientific

---

[36] This does not mean that abstraction is absent in Ao philosophy but that abstraction is insufficient for explaining a whole range of questions that affect one's experience of the Good.

[37] Sashinungla, 2021b, 94–95

[38] Within a tribal connectional metaphysics, this knowing (something) comes via understanding of the web of relations in which it participates. Knowing is always hinged upon knowing its relation to the whole.

view of "consciousness" as localized to individual minds, the tribal elemental tradition adopts a connectional, non-local view of *consciousness*—in conceiving of the spaces-between-people as being full of communications and relationships, and human selves as potential receivers and participants to those communications and interactions. This inherent principle of non-locality within indigenous cosmologies is shared by many such communities around the world.[39]

On the whole, the tribal systems of thought looked at knowledge largely as an experience, as "knowing," an internalized acquisition, or to use a Greek term as "gnosis." In contrast, the western system thinks of knowledge as a known, as "logos." They may intersect, but can hardly fit together in any single frame. Tribal cosmologies leave room for more ontological possibility than the dominant cosmological frameworks. Too much importance to a dogmatic reason and rationality resulted in the unfortunate loss of a sense of transcendence and of the value of life (human and natural). For too many years, we got used to the narrative that we are above the ecological system. It is ironic to call ourselves "wise humans" when we are undermining the very basis of our survival and that of so many other animals and plants. If we are to overcome our impending socio-ecological *crisis*,[40] then: in the individual level, we need to reconnect to our own bodies. People live too much in their mind. So, reconnect to the body of the world. In the conceptual/institutional level, we need new and novel ways of thinking. We need new stories–*new cosmologies*[41] about our place in the world in order to bring into alignment the complex contradictions required for maintaining our shared and diverse communities.

No model proves the ontological nonexistence of anything that the model's ontology excludes. The truth is, our adherence to a model highly determines the way we think and prevents us from thinking of anything that the model does not commit us to accept. We have long given up Cartesian dualism and scientism, for which subjects are the sole agents and objects are bereft of all agency—unless we are too caught up in and biased by the rationalist dogma of a self-sufficient, autonomous reason. We have crossed over from the sureties of Modernity to the

[39] Quantum physicists and philosophy of consciousness researchers, who pursue non-local ideas of consciousness are carrying forward this perspective. See David Bohm, *Wholeness and the Implicate Order* (London & New York: Routledge, 2002); David Bohm and Basil J. Hiley, *The Undivided Universe: An Ontological Interpretation of Quantum Theory* (London & New York: Routledge, 1993); David Bohm and Mark Edwards, *Changing Consciousness: Exploring the Hidden Source of the Social, Political, and environmental Crises Facing our World*, 1st ed. (San Francisco: Harper San Francisco, 1991); Amit Goswami, *The Physicists' View of Nature*, 2 vols. (New York: Kluwer Academic/Plenum Publishers, 2000); Amit Goswami, *God is not Dead: What Quantum Physics Tells Us About Our Origins and How We would Live* (Charlottesville: Hampton Roads, 2011); Amit Goswami, Richard E. Reed, and Maggie Goswami, *The Self-Aware Universe: How Consciousness Creates the Material World* (New York: Putnam, 1993).

[40] A crisis, according to deep ecologist Joanna Macy, which is at the root, created by human consciousness due to a dysfunctional view of ourselves as individualized and disconnected from others, and the greater environment. See Joanna Macy, "The Greening of the Self," in *Spiritual Ecology: The Cry of the Earth, a Collection of Essays*, edited by Llewellyn Vaughan-Lee (Point Reyes station: The Golden Sufi, 2013), 145–56.

[41] Long-standing, existing cosmologies but obscured by dominant frame and worldview.

fragmentations of Postmodernity. These fragmentations can be made sense of, not by reasserting some identities and their philosophies and rejecting others, but by articulating the meaning of all of them while continuously dialoguing with them critically, both by insiders and outsiders. Tribal philosophy contributes to this critical dialogue.

Bibliography

Ao, Tajen. 1980. *Ao Naga Customary Laws*. Mokokchung: Aowati Imchen.

Bohm, David. 2002. *Wholeness and the Implicate Order*. London/New York: Routledge.

Bohm, David, and Mark Edwards. 1991. *Changing Consciousness: Exploring the Hidden Source of the Social, Political, and environmental Crises Facing Our World*. 1st ed. San Francisco: Harper San Francisco.

Bohm, David, and Basil J. Hiley. 1993. *The Undivided Universe: An Ontological Interpretation of Quantum Theory*. London/New York: Routledge.

Cheney, Jim. 2005. Truth, Knowledge and the Wild World. *Ethics and the Environment* 10 (2): 105.

Deloria, Vine. 1979/2012. *The Metaphysics of Modern Existence*. San Francisco: Harper & Row.

Deloria, Vine, and Daniel R. Wildcat. 2001. *Power and place: Indian education in America*. Golden: Fulcrum Resources.

Goswami, Amit. 2000. *The Physicists' View of Nature*, 2 vols. New York: Kluwer Academic/Plenum Publishers.

———. 2011. *God is Not Dead: What Quantum Physics Tells Us About Our Origins and How We Would Live*. Charlottesville: Hampton Roads.

Goswami, Amit, Richard E. Reed, and Maggie Goswami. 1993. *The Self-Aware Universe: How Consciousness Creates the Material World*. New York: Putnam.

Hume, Lynne. 2002. *Ancestral power: The dreaming, consciousness, and Aboriginal Australians*. Carlton South: Melbourne University Press.

Imchen, Panger. 1993. *Ancient Ao Naga Religion and Culture*. New Delhi: Har-Anand Publications.

Jamir, N. Talitemjen. 1986. *Asen Kin Sobalibaren*. Kohima: N. Talitemjen Jamir.

Locke, John. 1980. In *Second Treatise of Government*, ed. C.B. Macpherson, reprint ed. Indianapolis: Hackett publishing company.

Macy, Joanna. 2013. The Greening of the Self. In *Spiritual Ecology: The Cry of the Earth, a Collection of Essays*, ed. Llewellyn Vaughan-Lee. Point Reyes Station: The Golden Sufi.

Maffie, James. 2014. *Aztec Philosophy: Understanding a World in Motion*. Boulder: University Press of Colorado.

Mills, J.P. 1973. *The Ao-Nagas*. London: Oxford University Press.

Miri, Sujata. 2001. *Ethics and Environment: Theory and the Adi Khasi Practice*. Guwahati/Delhi: Spectrum Publications.

Miri, Mrinal. 2011. The Spiritual and the Moral. In *The Oxford Anthology of Writings from North-East India: Poetry and Essays*, ed. Tilottoma Misra. Oxford: Oxford University Press.

Muecke, Stephen. 2004. *Ancient and Modern: Time, Culture and Indigenous Philosophy*. Sydney: UNSW Press.

Preston, C.J. 2005. Epistemology and Environmental Philosophy: The Epistemic Significance of Place. *Ethics and the Environment* 10 (2): 1–4.

Sashinungla. 2004. Ao Worldview. In *Rationality and Tribal Thought*, ed. Sujata Miri. New Delhi: Mittal Publication.

———. 2021. God-Self-World Continuum in Tribal Religion. *Poligrafi* 26 (103/104): 93–105.

———. 2021. Thinking, Looking and Acting: Beyond East and Southeast to the 'Other Asia'. In *The Interface of Domestic and International Factors in India's Foreign Policy*, ed. Johannes Dragsbaek Schmidt and Shantanu Chakrabarti. London/New York: Routledge.

Scott, David C. 1993. A mirror to M.M. Thomas's Perception on Inter-Religious Studies. *The SATHRI Journal* 2: 87–88.
Smith, Linda Tuhiwai. 2012. *Decolonizing methodologies: Research and indigenous peoples*. 2nd ed. London/New York: Zed Books.
Stewart, Georgina Tuari. 2021. *Maori Philosophy: Indigenous Thinking from Aotearoa*. London: Bloomsbury.
Waters, Anne. 2004. *American Indian Thought*. Oxford: Blackwell Publishing Ltd.

Sashinungla is Professor of Philosophy at Jadavpur University, Kolkata. Her main research interests are ethics, feminist philosophy, tribal philosophy, and environmental philosophy. Books include *Tradition and Modernity: Essays on Women of India*, ed. by Sashinungla and A.C. Sinha (Suryodaya Books, 2015), *Environment Preservation: A Philosophical Critique* (Decent Books, 2005), and *Ethics and Culture: Some Indian Reflection*, ed. by Sashinungla and I. Sanyal (Decent Books: 2010). She has published widely on feminist philosophy, insurgency, ethnic conflict, foreign policy, and tribal philosophy and culture with a particular focus on India's Northeast, and was Associate Editor of *Jadavpur Journal of Philosophy* from 2016–2021. She has also presented academic papers in several universities in the country and abroad. Current research projects focus on environmental humanities, tribal philosophy, and social and political ethics. **Personal Webpage**: https://jdvu.academia.edu/Sashinungla

Chapter 3
Elements of Slovenian Indigenous Religion

Lenart Škof

Abstract This essay is a philosophico-theological interpretation of the elements of Slovenian pre-Christian religious tradition called Staroverstvo (Nature Worshippers of Western Slovenia). This essay is based on Pavel Medveščček-Klančar's recent book *Iz nevidne strani neba* [*From the Invisible Side of the Sky*] and the first analysis of Slovenian pre-Christian religion as preserved in the Soča Valley until the late twentieth century and as documented in his book. Medveščček-Klančar's report and interpretation of this religious tradition is first compared to the contemporary theories of indigenous religions of the world. In the second part, we present our original philosophico-theological interpretation of the elements of this tradition in light of contemporary theories of process philosophy, quantum thinking in theology (i.e., entanglement), and horizontal transcendence within contemporary studies of natural religion.

Keywords Pre-Christian religion of Slovenia · Natural worshippers · Theology of nature · Slovenian indigenous religion · Elemental philosophical theology · Pavel Medveščček-Klančar

---

This chapter is a revised and abridged version of the chapter »God in Stone« from our book *God in Post-Christianity: An Elemental Philosophical Theology* (New York: SUNY Press, 2024). Reprinted with permission.

L. Škof (✉)
Institute for Philosophical and Religious Studies, Science and Research Centre Koper, Koper, Slovenia
e-mail: lenart.skof@guest.arnes.si

L. Škof et al. (eds.), *Elemental-Embodied Thinking for a New Era*, Sophia Studies in Cross-cultural Philosophy of Traditions and Cultures 42,
https://doi.org/10.1007/978-3-031-42119-8_3

On the Reconstruction of Slovenian Pre-Christian Religious Tradition

If we look into the synchronous study of contemporary philosophical theology, philosophy of religion and the cosmology of ancient religions, some quite fascinating questions are arising within some of the key areas of epistemology, ontology, and contemporary environmental ethics. The models of religious and theological experience encountered in this specific constellation range from the research into mystical and scientific correspondence between ancient and contemporary or new cosmologies (in humanities and sciences) to the interrogatives of new materialism within the framework of connections between philosophy, theology, and science. Perhaps the most vital part of new materialism in these constellations emerges as the very question of the *elemental* world of dwelling, around which most of our attention in this chapter will revolve. In this context, the elemental plane of dwelling is related to the awareness of a connection between beings (plants, animals, humans) and their ontological foundation that is revealed through the nature of the world that surrounds them—the meaning of the elemental world of dwelling is here linked to the old cosmological matrices of interdependence of dwelling in relation to the ancient "elements" of air, water, fire, and earth. The elemental character of the world of dwelling also embraces the mineral strata of nature (stones, metals, minerals), which are normally understood as inanimate or inorganic nature. Through elemental consciousness, nature unfolds as a space permeated by the omnipresent life force of all living things. In some cultures, this force is expressed through articulations about God; in others as the Supreme Being or life force; and in some this idea has been closely connected to nature.

Our chapter aims to show that in the human and social as well as natural and technical sciences (including the breakthrough findings of quantum physics in the early twentieth century), a confrontation between the original elemental cosmological matrices and contemporary theories has traced a space of ontological mystery and at the same time cosmological-quantum *entanglement*,[1] which both imply the possibility of transitioning to a newly conceived ontology and cosmotheology of man, nature and God. In this context, we will deal with extraordinary testimony of the Slovenian indigenous and pre-Christian religious tradition of *Staroverstvo* (called "Old Faith" but also known as *Naravoverstvo* or "Nature Worshippers"[2]),

[1] Catherine Keller in "The Entangled Cosmos: An Experiment in Physical Theopoetics" (*Journal of Cosmology* 20 [September 2012]: 8648–66) explains the phenomenon of entanglement of cosmos as follows: "The maximum has and yet might be called 'God.' But the physics of entanglement will not witness to the existence of a classical deity. It may however offer material evidence of a universe so mysteriously entangled as to escape the rival classicisms that pit science and theology against each other in the first place. What I want to share today belongs to the wider and polydoxical investigation of what I call 'apophatic entanglement.'" Thus, all phenomena in nature are ontologically entangled.

[2] *Staroverstvo* as used in our chapter is an exonym—namely, the members of this tradition have referred to themselves for the most of the time simply with the designations "us" and "ours." As

trying to provide it with a modern theological interpretation while abiding by the methods of new cosmology and new materialism. Of course, our designation of this religious tradition as an indigenous Slovenian religion requires an introductory note. In his chapter titled "Indigenous Lifeways and Knowing of the World," John Grim points out a stronger and a weaker definition of the notion of indigenous religion.[3] The former relates to the native population of an area that was colonised but has, nevertheless, retained some of its social, cultural, economic and political institutions; the latter, more specifically, relates to the concept of indigenousness as defined by these communities themselves, with regard to their histories, languages, cultures, beliefs, institutions, and territories as well as the related traditions. Grim defines the concept of indigenousness of a culture as that which refers to "small-scale societies around the planet who share and preserve ways of knowing the world embedded in particular languages, story cycles, kinship systems, world-view dispositions, and integrated relationships with the land on which they live."[4] Later on we will try to outline in what this tradition could be understood as an example of indigenous religion of Slovenia.

In 2015, an extraordinary testimony by Pavel Medveščck-Klančar (1933–2020) about the followers of Slovenian Nature Worshippers from the area of the Soča basin was published in the work *Iz nevidne strani neba* [*From the Invisible Side of the Sky*].[5] Before Medveščck, testimonies about this tradition from the area of the

"Old Faith" was also often derogatory employed towards them by the members of the Catholic Church, in this chapter, we will rather refer to their believers as "Nature Worshippers." See on this in Cirila Toplak, "Družbeno-politični vidiki zahodnoslovenske naravoverske skupnosti" ["Socio-political aspects of western Slovenian nature worshippers' community"], in: *Staroverstvo v Sloveniji med religijo in znanostjo* [*The Old Faith between Religion and Science*], edited by Saša Babič and Mateja Belak, *Studia mythologica Slavica*. Supplementa 17 (Ljubljana: Založba ZRC, 2022): 53f.

[3] For the strong definition, see The United Nations Declaration on the Rights of Indigenous Peoples (UNDRIP), adopted by the General Assembly on 13 September 2007.

[4] John Grim, "Indigenous Lifeways and Knowing of the World," in *The Oxford Handbook of Religion and Science*, eds. Philip Clayton and Zachary Simpson (Oxford: Oxford University Press, 2006), 87 f. Cf. here *Maori Philosophy* by Georgina Tuari Stewart (London: Bloomsbury, 2021), 3, which defines the fundamental characteristics of "Māori philosophy" as follows: "Past events do not lose their significance, and ancestors can collapse the space-time continuum to be co-present with their descendants" and "Knowledge in Māori terms is not restricted to the physical senses, but includes knowledge obtained through intuition and dreams."

[5] Pavel Medveščck-Klančar was from the Posočje region of Slovenia (i.e., the area of the Soča River Valley in the Western Slovenia). Apart from working as a teacher, writer, and graphic designer, Medveščck was also one of the most devoted collectors of Slovenian oral folk tradition. For over 30 years, he has been collecting oral testimonies from numerous informants. In this chapter, we refer to the oral testimonies of Janez Strgar (being his main and most important source and an authority of the tradition, and also Jože Pangerc, Jože Blažev, Toni Javor, Valentin Šmončev and Valentin Hvalica). The most important source among them was Janez Strgar. Cirila Toplak writes about Medveščck's ethnographic work as presented in his seminal book from 2015 and titled *Iz nevidne strani neba: razkrite skrivnosti staroverstva* [*From the Invisible Side of the Sky: Revealed Secrets of the Old Belief*] as follows: "On almost 600 pages, Pavel Medveščck revealed exclusive and well documented ethnographic records, mainly composed of interviews conducted with resi-

Slovenian Kras was published by Boris Čok.[6] Both works are based on the originally unwritten, i.e., oral tradition of Slovenian pre-Christian religion as it had been preserved through the centuries and passed on from generation to generation. Both books, especially Medveščeк's life's work, bring such rich messages and clues that they immediately invite direct comparison with certain key paradigms in the field of the very caesura between the oral and the written, and link to the issue of religious tradition reconstructions as broadly encompassed within the conceptual framework of ethnophilosophy.[7] Much like the efforts of contemporary researchers of autochthonous traditions of African peoples (and, by analogy, all other peoples of the world) which survived the era of colonisation and Christianisation and despite that managed to preserve the sacred lore of their ancestors, the investigation of the tradition of *Staroverstvo* also raises some critical questions—among which this treatise will highlight the importance of this tradition for contemporary philosophical

---

dents of remote hilly areas of Western Slovenia in the period from 1950 to 1978. The contents of these interviews could not be revealed earlier due to the oath of secrecy required of Pavel Medveščeк by his interviewees after they had confided in him and partially co-opted him into their secret community." About the tradition, Toplak also explains: "When discovered by Pavel Medveščeк in mid-twentieth century, the Nature Worshippers' community had been reduced to a few dozen elderly single male members, called uncles. According to their oral tradition, however, prior to the First World War at least several hundred people all together may still have been part of it in the Soča River valley. The existence of the community was concealed due to constant persecution perpetrated by the clergy and intolerant members of the Catholic majority. It was this outmost secrecy and geographic isolation in very remote subalpine and Dinaric areas of Western Slovenia as well as efficient leadership and organised defence against outer threats that decisively contributed to the survival of the community which, again according to their tradition, lasted centuries" (Cirila Toplak, "Tales in Social Practices of Nature Worshippers of Western Slovenia," *Acta histriae* 30 no. 3 [2022]: 628). For more on the Nature Worshippers' community and tradition, see Toplak's insightful paper.

[6] Pavel Medveščeк-Klančar, *Iz nevidne strani neba: razkrite skrivnosti staroverstva* [*From the Invisible Side of the Sky: Revealed Secrets of the Old Belief*] (Ljubljana: ZRS SAZU, 2015). This invaluable work was edited by Andrej Pleterski in Mateja Belak, with a scientific-critical study contributed by Andrej Pleterski. The volume was published in the collection Studia mythologica Slavica—Supplementa. In 2012, a testimony by Boris Čok titled *V siju mesečine. Ustno izročilo Lokve, Prelož in bližnje okolice* [*In the Glow of the Moon: Oral Heritage of the Villages of Lokev, Prelože and the Surrounding Areas*]. (Ljubljana: ZRC SAZU, 2012) was published in the same collection, describing the rich lore of the last representatives of Slovenian Nature Worshippers from the Kras region, another extraordinarily valuable cultural document about Slovenian indigenous religion (Kras is a limestone borderline plateau region in southwestern Slovenia).

[7] For a reconstruction of "African philosophy" (here in the broader sense of a set of mythological, religious, philosophical and ethnographic traditions) as ethnophilosophy, see Henry Odera Oruka, *Sage Philosophy: Indigenous Thinkers and Modern Debate on African Philosophy* (Leiden: Brill, 1990) and his essay "Four Trends in Current African Philosophy," in *The African Philosophy Reader*, eds. P.H. Coetzee and A.P.J. Roux (London: Routledge, 2003), 120–136. More on this in our treatise "Do Rta dobrega upanja in nazaj: tri vprašanja o afriški filozofiji," *Filozofski vestnik* 26, no. 3 (2005): 171–185, which, together with the related thematic block about African philosophy raises questions about the relationship between the oral and written traditions, about relationships between the mythologising (or ethnographically-centred) and rationalising (or analytical) currents in the studies of world traditions, and, ultimately, about the long-standing monocultural trend in the history of both Western philosophy and Christian theology.

theology and for the elemental and environmentally-oriented philosophy of religion. One of the world's greatest experts on the ancient oral traditions of the peoples of the world and the related environmental thought is American cultural ecologist and geophilosopher David Abram. When describing the lore of the Arctic Inuit and Yup'ik peoples in his essay "The Commonwealth of Breath," he approaches their centuries-old tradition and pays homage to it with these words:

> I shudder to speak of matters held so sacred to these oral traditions, and even more so to *write* of these understandings, which were never meant to be written down. [...] I bow to the various peoples indigenous to this region where I dwell, as well as to the animals, plants, and spirits of this parched terrain, asking their permission that I may write, here, of these things. Because these storied knowings, of which I comprehend so little, nonetheless need to be heard once again, and what is common among them needs to be felt, acknowledged, and replenished if life is to flourish.[8]

With this credo of humility and at the same time respect for tradition and life, Abram joins the ranks of those few thinkers—among them Medveščček—whose works reflect their authors' ability to lend an attentive and humble ear and to pass on the sacred traditions of old. They show the respect that had been lacking in an over a millennium-old position of arrogance and contempt from which others viewed these sacred traditions, derogatorily labelling them backward, pagan, pre-Christian and pre-scientific. Precisely the changes in contemporary humanistic and scientific movements brought the realisation that life on our planet will only thrive if we are able to understand these sacred old traditions in all their complexity and deep interconnectivity with nature and link them to the findings of modern science.

But before taking a closer look at the elements of Slovenian Nature Worshippers as they appear in our philosophical-theological consideration, there is another important point to note: the question whether this lore and the related religion could be considered as the indigenous religion in the ethnic territory of Slovenia—such was the self-understanding of certain believers from the Soča basin, as inferred from the words of one of them: "Most call it the old faith, but I call it *the first faith* that has survived, albeit in fragments, to this day."[9] Slovenian archaeologist, historian, and ethnologist Andrej Pleterski proposes two important hypotheses in this regard, which are related, respectively, to the narrower, Slavic, and the broader, Indo-European, religions. The first hypothesis relates to Nikolai Mikhailov's observation that "Slovenian mythology is the only Slavic tradition to have preserved so perfectly and thoroughly the so-called 'principal myth' of Slavic mythology," while in the second Pleterski assesses that "of all Indo-Europeans, the Slavs were the ones who led the simplest way of life the longest," and who "at the beginning of the Early Middle Ages were still living in conditions characteristic of the early Neolithic."[10]

---

[8] David Abram, "The Commonwealth of Breath," in *Atmospheres of Breathing*, eds. Lenart Škof and Petri Berndtson (New York: SUNY Press, 2018), 269.

[9] Vrhovčev Avguštin, 1962, as quoted in Medveščček-Klančar, *Iz nevidne strani neba,* 365 (our emphasis).

[10] See Chapter 6 titled "Verovanje host v sklopu staroverstva na Slovenskem in verovanja starih Slovanov" ["The belief of 'hoste' in the context of Old Belief in Slovenia and the belief of the

Pleterski bases his two premises on the preserved archaic structure in the myth of the god Triglav [The Three-Headed God], a tale learned from the Slovenian Nature Worshippers and written down by Boris Čok in his book *In the Glow of the Moon*. Within this context, Pleterski thus makes the assumption of a directly preserved archaic (or indigenous, as we wish to argue) tradition, which displays in its genealogy an *uninterrupted* (sic!) link that follows the axis of the Slavic mythological and religious traditions while also extending beyond the lore of Slavic Indo-European mythology as far back as the Palaeolithic. As we shall see, it is precisely in the elemental symbolism of the stone and related elements that manifests the secret connection to modern theology and even science which we attempt to trace and describe in this chapter. Both observations are important for us for the following reasons: firstly, they enable the assessment of a new theological reconstruction of the earliest traditions of humanity through their inherent and uninterrupted elemental matrix; secondly, they acknowledge *Staroverstvo* as an autochthonous or primordial, and thus—as we wish to contend on a basis of this argumentation—*indigenous religion* of Slovenian territories and bring it into direct dialogue with the neo-materialist and elemental trend of contemporary post-Christian theology in the making, to which our work belongs.[11]

In his introductory study to the volume *From the Invisible Side of the Sky*, Pleterski defines this tradition as the one cultural code that can help us gain new insight into the understanding of our common civilisation, and Medveščekʼs work as “unprecedented in Slovenia and outside it.”[12] Historically, this religion was hardly

ancient Slavs”] by Andrej Pleterski, in *Staroverstvo v Sloveniji med religijo in znanostjo*, 117. For the hypothesis by Nikolai Mikhailov, see *Zgodovina slovanske mitologije v XX. stoletju* [*The History of Slovenian Mythology in the 20th Century*] (Ljubljana: ZRC SAZU, 2021), 182. See in this regard also Andrej Pleterski, *Kulturni genom*: *prostor in njegovi ideogrami mitične zgodbe* [*The Cultural Genome: Space and its Ideograms of the Mythical Story*] (Ljubljana: ZRC SAZU, 2014), 76–85.

[11] The question whether it could be argued for the strong definition of the indigeneity of Slovenian Nature Worshippers shall remains open here, although *Staroverstvo* as a cultural, ethnic and religious tradition would hardly be able to pertain to its original scope: due to the socio-political, economic, and environmental causes, the meritocratic succession (as based on their political and religious leaders—called *dehnars*) was critically interrupted already after the WW2 and at least since the 80s or 90s Slovenian Nature Worshippers could be regarded as historical (I would still restrain from calling this tradition as extinct as we cannot know for certain whether some rites were still being preserved by the remaining members in secrecy until now). As argued by Toplak: “During the 20th century, the Nature Worshippers' community gradually disintegrated due to environmental destruction and human loss, especially that of their leaders dehnars, caused by both World Wars. Modernisation processes (infrastructure construction, industrialisation, and urbanisation) severed the inter-generational ties and depopulated remote rural areas.” (Toplak, “Tales in Social Practices of Nature Worshippers of Western Slovenia,” 631). As Medveššček (who unfortunately passed away in 2020) was admitted to the tradition (still, he was not a *dehnar*), he could then be regarded as (one of the) last Nature Worshipper(s). I thank Cirila Toplak for her clarifications of some of my conjectures.

[12] Medveššček-Klančar, *Iz nevidne strani neba*, 15. On the basis of exceptionally close connection of this religion with nature, Pleterski also refers to it with the equivalent term “Naturalism.” In the conclusion to his introductory study he writes: “It is precisely this 'inner voice' with its deep ethics

discussed at all, with its followers keeping it a strict secret, and when it *was* spoken about, it was by Christians, who did not understand it and pejoratively designated it pagan. We should therefore waken at Boris Čok's simple, yet deep and honest realisation (emphasised in his *In the Glow of the Moon*) that even as a child he could not embrace the images in the readings from the Bible (the desert, the strange clothes, unusual customs, remote myths…), for they felt alien or non-autochthonous to him at that time. Something completely different was listening to the stories told by his family, which, as he later understood, contained ancient and autochthonous elements of what nowadays can be called *Staroverstvo*.[13] As we shall see, this not only includes archaic mythological patterns and a variety of divine names, but also testifies to an extraordinary presence of gender distinction—i.e., the feminine and masculine poles of the divinity in the very core of this tradition, which was, almost without exception, obscured or suppressed in every major world religion.[14] Our task is to talk with respect about the remarkable value of this uninterrupted cultural history of Slovenia.

Elements of Slovenian Indigenous Religion

In his book *Power and Place*, theologian Vine Victor Deloria Jr. (1933–2005; also known as Standing Rock Sioux), from the Sioux tribe of American Indians, points out the fact that white men scorned the knowledge of American Indians for centuries, regarding their teachings as mere superstition and at the same time believing that "a complex mixture of folklore, religious doctrine, and Greek natural sciences, was the highest intellectual achievement of our species."[15] Only in recent decades has it become evident that American Indian tribes (and, by analogy, other tribes, small-scale societies, and cultures of the world) have been in possession of special and important knowledge about the cosmos and the natural world. The concepts that Deloria puts to the fore are power and place. While the dominant Western tradition in philosophy and science spoke and thought only through the coordinate system of linear space, time and energy at least up to Bergson or Whitehead, indigenous people have always spoken and thought through the matrix of power/life force and space. But their ideas about the common characteristics of, say, humans and

---

and infinite respect for nature and its balance that gives this book its incommensurable, unique and significant value, especially in this rampant globalised world that only cares about profit." Ibid., 31.

[13] Cf. Čok, *V siju mesečine*, 12f.

[14] See our *Antigone's Sisters: On the Matrix of Love* (Albany, NY: SUNY Press, 2021). Exceptions being some parts of Hinduism and Buddhism (particularly in the framework of Tantrism), the tradition of Sophia/Wisdom within the Judeo-Christian tradition and, of course, the currents of contemporary feminist theology. Another example where the feminine principle is also considered would definitely be philosophical Daoism in the context of Chinese tradition.

[15] Vine Deloria Jr. and Daniel Wildcat, *Power & Place: Indian Education in America* (Golden, CO: American Indian Graduate Center and Fulcrum Resources, 2001), 1.

animals, or even about the interconnectedness of the mineral, plant, animal and human worlds, did not fit into the conceptual framework of tradition nor could this knowledge be framed within the domain of the term "faith." Deloria calls attention to another important dimension that will interest us in our treatise, when he says: "We may grant that the energy described by quantum physics appears to be identical to the mysterious power that almost all tribes accepted as the primary constituent of the universe."[16] The ultimate realisation of the philosophical theology of American Indians, as presented by Deloria, that our dwelling is essentially *relational* and connected to other living beings, the land and its elements, opens up ample possibilities of reinterpretation of numerous similar spiritual traditions—of these we shall foreground the philosophical-theological view of the Slovenian indigenous religion.[17]

The elemental philosophical theology developed in this treatise is an expression of hidden or (still) mysterious cosmic correspondences between different types of living beings that inhabit and fill the space. As the Nature Worshippers of Slovenia put into the fore stones and their inherent *spiritual materiality*, we shall start with them. In the jargon of modern quantum physics, Catherine Keller describes the ontological status of a stone as follows:

> In *Process and Reality*, Whitehead would formulate this radically nonsimple location as his "ontological principle": "Everything is positively somewhere in actuality and in potency everywhere." This potentiality, suggestive of the quantum potential or vacuum, also anticipates the French quantum physicist Bernard d'Espagnat's snapshot of a stone: "its 'quantum state' is 'entangled' (this is the technical word) with the state of the whole Universe."[18]

---

[16] Ibid., 5.

[17] The term "spiritual" used in this treatise refers to the elemental presence of life energy in us which, unlike the classical metaphysical and scientific views of "spirituality," is not understood as a Platonist-Cartesian antipode to the body, but quite the opposite, as a vital force (i.e., *spiritual matter*) that is in its nature connected to air, breath and the related vital bodily energies of the (meso)cosmos, nature and mankind. See about this Lenart Škof, *Breath of Proximity: Intersubjectivity, Ethics and Peace* (Dordrecht: Springer, 2015).

[18] Catherine Keller, "Tingles of Matter, Tangles of Theology," in *Entangled Worlds: Religion, Science, and New Materialism*, eds. Catherine Keller and Mary-Jane Rubenstein (New York: Fordham University Press, 2017), 118f.

An interesting analogy with regard to the mineral core and stones can be found in a work by Ernst Haeckel, a German zoologist and philosopher of biology. In his final work, titled *Kristallseelen: Studien über das anorganische Leben* (accessible in a 1917 edition by Pranava Books), Haeckel developed an idiosyncratic and at the time utterly uncomprehended theory of the "psyche" of minerals, stating the following: "Alle Substanz besitzt Leben, anorganische ebenso wie organische; alle Dinge sind beseelt, Kristalle so gut wie Organismen." (viii) Cf. also his penultimate work *Gott-Natur (Theophysis)* (Leipzig: Alfred Kröner Verlag, 1914), in which he put forth a monistic interpretation of God, i.e., as expressing itself through all nature. Lamentably, it should also be pointed out that Haeckel, as an evolutionist, defended extremist eugenic and racist positions, thereby simultaneously and inherently refuting his own theory about the theological-ontological intertwinement of all living beings.

Medvešček's book, at which we are going to take a closer look now,[19] contains remarkable testimony to the intertwinement or ontological entanglement of everything that exists on Earth. An extraordinary example of such understanding is the testimony by Janez Strgar, who ushered Medvešček into the sacred realm of his faith. This is what he said about the faith of Toni Javor, the only one, according to Strgar, among his friends who had frequent contacts with the other world:

> On such days he would take time to himself, spend it alone in the mountain or in the woods where he could see, as he put it, beyond the visible, where one could hear echoes and voices from primeval time. [...] Javor also refused to accept the concept of inanimate matter, for to him everything that surrounded him was living matter. [...] They [*i.e., his younger contemporaries*] find it particularly silly when Javor says he is also friends with trees, water and certain stones. In the end, there follows the teaching that says we should love nature like a mother, for we are part of it. (56)

This passage somehow brings together all that we intend to think about here: the Nature Worshippers were able to cross the linearity of time via their solitary, sometimes even truly ascetic detachment, and enter the so-called timelessness in which they felt the interconnection of all existing things as animate—i.e., of the dynamic whole of spiritual matter and the supporting primal force of Nikrmana (The Creatrix) in one of its emanations (it was sighted during the nights when lightning would strike; some saw it in the form of a woman, others as clouds, or also in the form of a bull, a snake, an ibex, or a horse. Nikrmana is an indication that the whole space was revered for the purposes determined by this primal force—this is the clear link connecting the Slovenian indigenous religion to other indigenous religions from elsewhere in the world. Based on the tradition and secret teachings passed on to him, Medvešček is aware that "a stone is not just inanimate matter, but an important element in space and a human's companion that co-creates and ensures human existence" (525). Stones resembling *a snake's head* selected by Nature Worshippers were also worshipped as protectors of seeds, or, in the words of Valentin Hvalica: "The great stone acted as an altar to which people offered the precious seeds on which they depended. In doing so, they also helped the birds, which are a living link between the Earth and the sky" (235).

According to ancient belief, stones symbolized all living nature, and taking into account the previously mentioned credo of quantum physicist Bernard d'Espagnat that the quantum status of a stone is "entangled" with the status of the whole universe, we can derive the conclusion that there existed special knowledge that was for centuries available to Slovenian Nature Worshippers who saw and felt in the consecrated stones an extraordinary power or force— moreover, Valentin Hvalica even says that "when you 'rock' the *green snake head* with some special purpose you may hear the primal time, for the green snake head was born in the sacred river that springs from the heart of the sacred mountain of Triglav, in which our belief is rooted" (236).

---

[19] All page number references for the cited passages are from Medvešček-Klančar, *Iz nevidne strani neba.*

Stones were also the composing elements of the *tročan* or triad/trinity, which connected in itself the elements of earth, water and fire. When building a house, Nature Worshippers selected three stones and laid them so as to form a sacred triangle to express this interconnectedness. Such a *tročan* was believed to protect the house and have a beneficial effect on humans, animals, and plants. It kept the world in balance by way of special cosmic correlation and the knowledge that made that possible was a carefully guarded secret that was passed from father to son, and only in the hour of death. A *tročan* also had an ethical meaning. When Janez Strgar spoke of the *tročan*, he called attention to its cosmic-ethical dimension:

> A *tročan* is also something that exists within ourselves, though not inside a single person, it's something which at present exists between you, Tone and me. We cannot be alone, one without the other. Only then does one feel this inner power within oneself. (87–88)

In the context of *Staroverstvo*, this view demonstrates the extraordinary belief (which lies at the foundation of modern environmental ethics) "that in nature, we are all equally important and interdependent" (108). *Tročans* point to a very important characteristic now: namely, that its view of relationships in the world is a matter of *horizontal transcendence*—both at the level of relationships with the macrocosm and at the level of relationships between living beings as parts of nature. Conversely, the classical theology of monotheist religions insisted on vertical or monarchic transcendence, which denied the people (particularly women and children) their full ontological affiliation and did not grant any ontological rights to animals and plants, let alone so-called inanimate nature. As such it was too often intolerant of and violent towards the allegedly pagan religions that exhibited that, including—unfortunately—the Nature Worshippers, which it did not understand. Finally, there is also the blood *tročan* connecting the mother, the father and the child. This triad also exhibits horizontal genealogy in a special genealogical relationship established between the parents and the children, wherein the child as a part of this triad can now exist as an ontologically equal element of the structure. This is also one of the reasons why the tradition of *Staroverstvo* prohibits violence against children (as opposed to Christianity, which has supported it for centuries and, lamentably, preserved it to this day).[20]

To recap—we can now say that in the cosmic-theological sense, it is the elemental horizontal ontological connection that enables the understanding of this secret (in the sense of secret cosmic correspondence) connection between the human and nature and its forces—as Valentin Šmončev beautifully described when he said that one should live in such a way as "to equate oneself with all living beings living around us and with us" and that "there is also among us that which we cannot see or

[20] Jože Pangerc thus says that unlike "teachers and preachers, who punish children by making them kneel on hard grain for hours," Old Faith believers do not subject their children to corporal punishment; they follow nature in this too, for animals never torture their young (Medveščck-Klančar, *Iz nevidne strani neba*, 463). This remarkable note reflects the respect for children, which has been absent throughout histories of philosophical thought and theology. On the other hand, the symptom of the Christian attitude towards children and of a lacking ontology of the child manifests in the uncountable instances of sexual abuse of children and adolescents in the Catholic Church.

understand but which definitely works" (179). At this point we may recall Deloria saying that the energy described by quantum physics appears to be that secret force that ancient peoples and traditions considered to be the main component of the universe. An extraordinary testimony to this mysterious force is provided by Jože Fatorinov, to whom this force revealed while he was observing a pool in the river. Following a poetic description of the mystical experience he was privy to he writes:

> A secret force awoke in me, filling up my body and making it increasingly unrecognisable, different. I felt an amazing power in me. Since then I haven't needed anything else. Everything I need is within arm's reach. But it is up to me to continue to be honest and good and accept all that my friends-uncles do. (192)

It is today amazing how highly Nature Worshippers thought of nature. It was the cathedral of their "Church," and the two greatest sanctuaries of Babja jama [*The cave of old woman*] and Padence (an area close to Most na Soči) were only spoken of with deepest reverence. Their insights and teachings about nature are in perfect keeping with the latest insights of environmental humanities and environmental ethics. Perhaps the most compelling testimony on this topic in the entire Medveščček's book is that by his informant Jože Blažev. I shall recap it here by way of introduction to the reflection about the importance of new elemental philosophical theology for the modern debate on the environment. This is what he tells Medveščček in an extraordinary passage related to the devastation of war and arrival of industry in the Soča Valley, a valley so sacred to Nature Worshippers:

> You know what, they destroyed everything. Even the great *tročan*. It was a really large *tročan* formed by [the mountain of] Jelenk, the Jazbenk [chasm] and the Babja jama [cave]. All three were connected to the netherworld, where the great white snake lived. When there was high water next to the cave, it was a sign that the snake did not want anyone to come close. It longed for peace and quiet … […] This vulture [i.e., modern man] is destroying everything and will in the end destroy himself too. That will be a relief for nature, which will be able to recover with the help of Nikrmana. Of course, I won't live to see that day and neither will you. (441)

Nature Worshippers were also called *kačarji* [worshippers of a serpent: *kača* in Slovenian] and one can only imagine the scorn felt for this faith by Christians, who had been instilled by their tradition with hatred for an animal so sacred to them. Their myth of the creation of life says there were once four moons in the sky: the largest was called the procreation moon, and when it cracked, seeds and eggs seeped from its cracks and fell on the Earth. From the first egg hatched a *white snake*, the symbol of *Staroverstvo*. It settled in Babja jama, making this cave the central Nature Worshippers' sanctuary. All other animals hatched from its eggs there, and all plants sprouted from the seeds. Only then did humanity emerge—from the procreation moon (which had in the meantime become a star), broken in half and fallen onto the Earth, developed the first woman and the first man as the two halves of that moon. This exceptional genealogy suggests a cosmological origin of all living beings, which is inconsistent with the theological and monarchic-monotheistic myth from the Judeo-Christian tradition: "In that day the LORD with his hard and great and strong sword will punish Leviathan the fleeing serpent, Leviathan the twisting serpent, and he will slay the dragon that is in the sea" (Is 27:1). Of course, one should not forget the ancient Indo-European myth about the snake-defeating hero (in the old Indian myth regarding the Vedas, for example, that hero is Indra, who overcomes the cosmic obstacle in the form of snake/dragon called Vritra or the one who kept the waters of the world captive and therewith life; cf. also later Christian versions from the stories about Saint George slaying a dragon), and Pleterski, too, mentions this aspect of the mythical story related to the mytheme of the snake.[21] The Nature Worshippers' myth of the snake thus suggests this is an extremely old subject that does not depart from the elemental world of dwelling (with the elements of water, fire, and air, as well as earth—including the netherworld) and continues to maintain some kind of archaic (pre-Biblical and pre-Christian and preceding later Indo-European corrections towards establishing a new and allegedly higher moral order by killing a primordial mythological snake/dragon) elemental awareness about the world and the life in it. But how do these elemental strata of dwelling manifest? How did the Nature Worshippers link their faith to them?

The world we inhabit is thus divided into the netherworld, the earthly world and the heavenly world. All three worlds stay in balance through coordinated "cooperation" (or mystic/secret cosmic correspondences) of the beings of nature that have inhabited these worlds—divinities, humans, animals, plants, and rocks/stones. In the sense of horizontal transcendence, an invisible, yet sacred connection is preserved among the beings of the world, which is strengthened by the belief that the spirits of the deceased settle as *zduhci* (plural of *zduhec*, meaning 'wind spirit') in other beings and remain in contact with us even after death—every Nature Worshipper had their own *zduhec* in some living being through which they maintained this contact with the other or, better, parallel world. This connection is essentially a bond of love, as it contains no elements of violence, hatred, cruelty, or deceit. Through communication with *zduhci,* they travelled into timelessness, in which they were able to establish contact with the memories, suffering and experiences of their ancestors. It follows that crushing a bug or a butterfly could mean killing the one "who lives in them their second life" (205). Their ethics is in this aspect absolutely comparable with the Jainist ethics of respect and preservation of all living beings, the strictest among the ethics of respect for life.

The Earth that we inhabit has been soaked by blood and pervaded by pain, terror and suffering for millennia. The industrial revolution and an unsuspected technological development transformed it into something readily available for human use and abuse. Man has diverted, restricted or stopped the waters, polluting them and making many unfit for life; he has tamed the fire and transformed it into an energy of unimagined force, down to its nuclear and destructive power; he has turned the

[21] Medveščček-Klančar, *Iz nevidne strani neba*, 23 (Introduction by A. Pleterski). For the dragon motif, see Vladimir Yakovlevich Propp, *Zgodovinske korenine čarobne pravljice* [*The Historical Roots of Fairy-tales*] (Ljubljana: ZRC SAZU: Inštitut za slovensko narodopisje, 2013), 191–246. Initially, Propp links the dragon motif to the elements—particularly fire and water. Later he also refers to ancient Babylonian tradition: "In the Babylonian creation myth it is said: 'When Tiamat opened her mouth to its full extent, he [Marduk] let in the Evil Wind [Imhullu] so that she close not her lips" (Ibid., 213).

air into a weapon, a medium for poisoning, choking and killing other living beings, and riddled the earth with remains of weapons, of metal shaped into bullets and missiles. Man has blasted numerous sacred stones and rocks of Nature Worshippers, destroying the old *tročans* that used to preserve the balance of the believers' dwelling in the world, of passages from one season into the other, from life to death and into new life again. Few individuals and even fewer traditions have found the strength to oppose that: the Slovenian Nature Worshippers are undoubtedly a remarkable testimony to that, one that survived till the end of the twentieth century.

Appendix: Manifesto for a Free Earth

by Marko Pogačnik[22]

Mother, Mother, is it true that "climate change" will turn green meadows into wastelands and dry up the clear rivers and streams?
Mother, Mother, is it true that masses of climate refugees will flood Europe and famine will plague human populations between the South and the North?
Mother, Mother, is it true that the Sun will shine so brightly that the ladybirds will fly away and people will go to the afterlife in huge waves?
Mother, Mother, is the Earth really a dead thing, incapable of helping animals, plants, people, landscapes and dolphins in dire need?

The Earth is not a dead thing made up of hard rocks, clothed only in a thin layer of life tissue. The Earth is first and foremost an elemental consciousness.
The Earth as a conscious being knows that the cosmic cycle has turned, that the time of human autocracy inflated by the will to power and authority has come to an end.
The Earth as a co-creator in the revelation of a new round of cosmic evolution, knows that an age of interplay of opposites is coming.
The Earth is a living being made up of countless tiny suns, which hold a memory from eternity to the end of eternity.

In order to prevent the disintegration of the fabric of life, the Earth has raised the energetic pulse of its worlds and their beings to a new level of existence.

[22] Marko Pogačnik (1944) lives with his wife and collaborator Marika in Šempas, Slovenia. During the years 1965–71 he worked in Conceptual Art and Land Art as member of OHO group. After 1971 he works in the field of art combined with integral ecology (geomancy). He has developed a method of Earth healing called "lithopuncture" with stones standing on specific points and complemented with carved cosmograms. He developed "Gaia Touch" body exercises to tune to the essence of the Earth. In 2006 he started the world-wide project of Geopuncture Circles. His books among other: *Gaia's Quantum Leap, Universe of the Human Body, Christ Power and the Earth Wisdom, Dancing with the Earth Changes, Creating Gaia Culture*. In 2016, he was appointed UNESCO Artist for Peace (for the period 2016–2022). www.markopogacnik.com

In order to provide enough space for all beings, visible and invisible, to evolve, the Earth is opening the door to spaces that are currently unknown to us.
Split into different dimensions, the Earth is able to accommodate all beings determined to preserve and love the Earth as a place of creative coexistence.
At the same time, the Earth closes the door to forces and powers that are not dedicated to life and are not ready to exist and act in harmony with the essence of being.

The Earth, as the creator of the embodied world, opens the gates of the Water element, leading into the halls of health and the landscapes of healing.
The Earth, as the mother of embodied creation, opens the gates of the Earth element, leading to the causal worlds of being where ideas take form.
The Earth, as the creator of reality, opens the gates of the Fire element, leading to truth as the simple truth of life.
The Earth, as the mother of the world of revelation, opens the gates of the Air element, leading to the insight of what is the meaning of existence and creation on Earth.

Mother, Mother, is it true that the landscapes of the Earth are inhabited by innumerable elemental beings invisible to us, the guardians of life in nature?
Mother, Mother, is it true that the dead dwell not far from us as invisible companions and guardians of the incarnate human race?
Mother, Mother, is it true that the dragons sustain and renew the core force of life that cannot be erased, suppressed or permanently enslaved?
Mother, Mother, is the Earth really a living expression of infinite love, capable of realising the paradisiacal value of being if we choose to listen her inspiration?

November 2022

Bibliography

Abram, David. 2018. The Commonwealth of Breath. In *Atmospheres of Breathing*, ed. Lenart Škof and Petri Berndtson, 263–276. New York: SUNY Press.

Čok, Boris. 2012. *V siju mesečine: ustno izročilo Lokve, Prelož in bližnje okolice*. Ljubljana: ZRC SAZU.

Deloria, Vine, Jr., and Daniel Wildcat. 2001. *Power & Place: Indian Education in America*. Golden: American Indian Graduate Center and Fulcrum Resources.

Grim, John. 2006. Indigenous Lifeways and Knowing of the World. In *The Oxford Handbook of Religion and Science*, ed. Philip Clayton and Zachary Simpson, 87–107. Oxford: Oxford University Press.

Haeckel, Ernst. 1914. *Gott-Natur (Theophysis)*. Leipzig: Alfred Kröner Verlag.

———. 1917. *Kristallseelen: Studien über das anorganische Leben*. Delhi: Pranava Books.

Keller, Catherine. 2012. The Entangled Cosmos: An Experiment in Physical Theopoetics. *Journal of Cosmology* 20: 8648–8666.

———. 2017. Tingles of Matter, Tangles of Theology. In *Entangled Worlds: Religion, Science, and New Materialism*, ed. Catherine Keller and Mary-Jane Rubenstein, 111–135. New York: Fordham University Press.

Medveščck-Klančar, Pavel. 2015. *Iz nevidne strani neba: razkrite skrivnosti staroverstva*. Ljubljana: ZRC SAZU.

Oruka, Odera, and Henry. 1990. *Sage Philosophy: Indigenous Thinkers and Modern Debate on African Philosophy*. Leiden: Brill.

———. 2003. Four Trends in Current African Philosophy. In *The African Philosophy Reader*, ed. P.H. Coetzee and A.P.J. Roux, 120–136. London: Routledge.

Pleterski, Andrej. 2014. *Kulturni genom: prostor in njegovi ideogrami mitične zgodbe*. Ljubljana: ZRC SAZU.

———. 2022. Verovanje host v sklopu staroverstva na Slovenskem in verovanja starih Slovanov. In *Staroverstvo v Sloveniji med religijo in znanostjo*, ed. Saša Babič and Mateja Belak, 109–143. Ljubljana: Založba ZRC. (Studia mythologica Slavica. Supplementa 17).

Propp, Vladimir Yakovlevich. 2013. *Zgodovinske korenine čarobne pravljice*. Ljubljana: ZRC SAZU: Inštitut za slovensko narodopisje.

Škof, Lenart. 2005. Do Rta dobrega upanja in nazaj: tri vprašanja o afriški filozofiji. *Filozofski vestnik* 26 (3): 171–185.

———. 2015. *Breath of Proximity: Intersubjectivity, Ethics and Peace*. Dordrecht: Springer.

———. 2021. *Antigone's Sisters: On the Matrix of Love*. Albany: SUNY Press.

Tuari Stewart, Georgina. 2021. *Maori Philosophy*. London: Bloomsbury.

Toplak, Cirila. 2022. Tales in Social Practices of Nature Worshippers of Western Slovenia. *Acta histriae* 30 (3): 627–654.

Lenart Škof is Head of the Institute for Philosophical and Religious Studies at the Science and Research Centre Koper (Koper, Slovenia) and Dean at Alma Mater Europaea—Institutum Studiorum Humanitatis (Ljubljana, Slovenia). He authored *Antigone's Sisters: On the Matrix of Love* (SUNY Press, 2021), *Breath of Proximity* (Springer, 2015) and *Pragmatist Variations on Ethical and Intercultural Life* (Lexington Books, 2012) and recently (co)edited *Shame, Gender Violence and Ethics: Terrors of Injustice* (Lexington Books, 2021) and *Atmospheres of Breathing* (SUNY Press, 2018). Dr. Škof is president of the Slovenian Society for Comparative Religion and member of the European Academy or Sciences and Arts. His main areas of research are elemental and respiratory philosophy, new cosmology, and philosophical theology. The title of his forthcoming book is *God in Post-Christianity: An Elemental Philosophical Theology.* Homepage: https://zrs-kp.academia.edu/LenartŠkof

Chapter 4
Experiencing Elemental Embodiment: (Re) visiting Latvian Folklore on Life and Nature

Anne Sauka

Abstract Latest research in environmental humanities often presumes the necessity of some kind of an ontological "shift" in thinking and living, while the question of the possibility of such a shift on an experiential level, is still to be answered. In this article, I am (re) visiting Latvian folk epistemologies as a sample case of alternate yet already "present" ontogenealogies that could be applied for reinventing ways to experience environed embodiment. While it is not possible, or desirable to recapture the past, alternate ontogenealogies of the Global North—partly local, yet always also entangled with closer and farther regions, can provide insight into the building of future societies upon the idea of meaningful, senseful materiality. Especially—in context of pre-Christian sacrality of naturecultures that place meaning beyond and before human I-consciousness, thus, capturing the idea of consciousness as the result, not the source of senseful agency of materiality. Moreover, such inquiry allows diluting the presumption of a homogeneous Global North, instead demonstrating the heterogenous genealogical background that co-constitutes our experiences and materialities today. Within this article, I focus on the topics of life and nature in Latvian folksongs—*Dainas,* expressed through the elemental worldings of stone and forest, to demonstrate old Latvian epistemologies as a fruitful ground for an affirmative environmental ethics in the future.

Keywords Latvian folklore · Ontogenealogies · Mythologies · Stone · Forest · Life · Nature · Environmental philosophy

A. Sauka (✉)
Department of Philosophy and Ethics, University of Latvia, Rīga, Latvia

L. Škof et al. (eds.), *Elemental-Embodied Thinking for a New Era*, Sophia Studies in Cross-cultural Philosophy of Traditions and Cultures 42,
https://doi.org/10.1007/978-3-031-42119-8_4

Stepping on the Path of Alternate Worldings for the Future

Latest research in humanities, particularly, in new materialist and posthumanist studies[1] either explicitly or implicitly often proposes an ontological shift in thinking and living as necessary for achieving ethical societal transformation in environmental contexts, and consequentially also a transformation of lived materialities for more-than-human futures.

In my research, I follow the idea that an ontological shift is only possible where it is thought of as part of already existing worldings.[2] I highlight the idea that such a shift is possible only there, where it relies on already pre-existing epistemologies or affectual human-environment relations that can be thought of as experiential, to the extent in which they have partaken and continue to partake in the materialities we live by. If these alternate experiential forms are to be found, they are to be sought for in the already present, yet alternate imaginaries[3] and worldings that are concealed or less represented within the homogenized, globalized narratives of the Global North.

Recognizing the entanglement of materialities and epistemologies, I conceptualize these worldings via the concept of ontogenealogies. With this concept, I propose to reinvent the Foucauldian idea of genealogical research within a material context,[4] where life itself is conceived as genealogical,[5] and to then conceptualize the genealogical entanglement of materialities and imaginaries.

I propose the assumption that to step on the path of alternate ontogenealogies, two different approaches are equally important:

1. a phenomenological or another kind of first-person approach (i.e., for example, embodied critical thinking)[6] that allows affectual experience of processuality,

[1] See Stacy Alaimo, *Bodily Natures: Science, Environment and the Material Self (*Bloomington: Indiana University Press, 2010), Astrida Neimanis, *Bodies of Water* (London: Bloomsbury, 2017). Esp. Astrida, Neimanis, Astrida Neimanis, "Posthuman Phenomenologies for Planetary Bodies of Water," in *A Feminist Companion to the Posthumanities*, ed. Cecilia Åsberg, Rosi Braidotti (Cham: Springer International Publishing AG, 2018), 55–66.

[2] "What is needed is a posthumanist ethics, an ethics of worlding" Karen Barad, *Meeting the Universe Halfway* (Durham: Duke University Press, 2007), 392.

[3] Marietta Radomska, *Uncontainable Life: A Biophilosophy of Bioart* (Linköping: Linköping University Electronic Press: 2016).

[4] Taking in mind the restrictions of space, I here omit some of the methodological considerations that do not refer directly to the topic at hand—however, since these are already discussed in some of my previous articles, I here cite some of the material that explores the topic of genealogy and ontogenealogy further. I first refer to the idea of reframing the sphere of critical genealogy from a methodological approach to a materially embedded ontological outlook in Anne Sauka, "The Nature of Our Becoming: Genealogical Perspectives," *Le foucaldien* 6, no. 1 (2020): 1–30.

[5] See, Anne Sauka, "Life in Process: The Lived-Body Ethics for Future", *Religious-Philosophical Articles* [Reliģiski—filozofiski raksti], 28 (2020): 154–183.

[6] The experiential outlook, with the highlight of the "already present" in the phenomenological perception, felt sense, and affectual relations with what is often termed as "nature" is taken on by the field of embodied critical thinking. Donata Schoeller and Sigridur Thorgeirsdottir, "Embodied

intercarnality, and the emergence of thought *before* I-consciousness,[7] thus, demonstrating the presence of embeddedness and transcorporeality in human day-to-day experiences, and
2. a genealogical revisitation and visitation of alternate and old epistemologies that bring about our worldings, while their presence might be largely unrecognized and concealed by the predominant genealogies of disembodiment and human detachment from nature and its elemental connectedness.

While previously I have dealt with these research approaches separately, here, I will draw together these two research lines via (re)visiting Latvian folk epistemologies as a sample case of alternate yet already "present" ontogenealogies that could be applied for reinventing ways to experience environed embodiment and to search for affirmative environmental ethicalities for the future. This idea of reconnection via revisitation is also reflected in Robin Wall Krimmerer's research, where she comments on her students' inability to describe any positive interactions between people and the land:

> As the land becomes impoverished, so too does the scope of their vision [...] How can we begin to move toward ecological and cultural sustainability if we cannot even imagine what the path feels like? [...] These students were not raised on the story of Skywoman.[8]

With this quote, Krimmerer succinctly brings together the necessity for an ontological shift, with the idea of revisiting old epistemologies that co-constitute us and our affectual relations with our environments, yet often remain hidden from plain sight. While indigenous knowledges, mostly from the Global South, but also including Saami and Inuit knowledge have been to some extent already reflected upon today[9] searching for valuable epistemologies for relating to the world in a more sustainable way that differs from the dominating homogenized Global Northern genealogies, I propose the significance of looking into our own—European pre-Christian epistemologies and recognizing how they remain co-constitutive for lived materialities today. This is a view that I share with authors such as Sophie Strand,[10] Nancie Marie

---

Critical Thinking: The Experiential Turn and its Transformative Aspects," *philoSOPHIA: A Journal of Continental Feminism*, 9 no. 1 (2019): 92–109. Schoeller and Thorgisdottir offer the return to nature through an experiential approach and offer "felt sensing" as a possible vehicle to reconnect with one's environment.

[7] Anne Sauka, "Beyond the Skin Line: Tuning into the Body-Environment. A Venture into the Before of Conceptualizations," *The Polish Journal of Aesthetics. Listening and Polyphony. Philosophy, Aesthetics*, 64, no.1 (2022): 162–181.

[8] Robin Krimmerer, *Braiding Sweetgrass* (Minneapolis: Milkweed Editions, 2013), *accessed online*.

[9] Here, next to Robin Wall Krimmerer, *Braiding Sweetgrass*, Tyson Yunkaporta, *Sand Talk: How Indigenous Thinking Can Save the World* (New York: HarperCollins, 2020) should also be mentioned.

[10] Sophie Strand, *The Flowering Wand* (Rochester: Inner Traditions, 2022).

Brown,[11] Lenart Škof[12] and Rune Hjarnø Rasmussen[13] who invite rethinking Global Northern, Icelandic, Slovenian and Old Norse mythologies/epistemologies, respectively, for relating to the world more sustainably today, thus continuing the turn towards (new) animism[14] in a Western/Global Northern context—not to appropriate indigenous knowledge, but rather to demonstrate the ontogenealogical heterogeneity[15] of the spaces largely thought of as homogeneous and summarized under the umbrella term "the Global North."[16]

While it is not possible, or desirable to recapture the past, alternate ontogenealogies of the Global North—partly local, yet always also entangled with closer and farther regions, provide insight into building future societies upon the idea of meaningful, senseful materiality, especially in the context of the pre-Christian sacrality of naturecultures that place meaning beyond and before human I-consciousness, thus, capturing the idea of consciousness as the result, not the source of senseful agency of materiality.

For this, Latvia and the Baltics are good examples as societies that are Christianized later than most others in the Global North (circa twentieth–fourteenth centuries),[17] and still retains parts of its "pagan" worldings. Embodied critical thinking (ECT) allows reflecting the examples from past epistemologies as experienced "instances" to relate to alternate ontogenealogies that co-constitute experiences while being concealed by the homogenized Global Northern narratives.

The "sacred scriptures" in Latvia are called "Daina"[18] and it is a form of traditional poetry that encompasses many subjects, from everyday life and work,

[11] Marie Nancy Brown, *Looking for the Hidden Folk* (New York: Pegasus Books, 2022).

[12] See Lenart Škof, "Elements of Slovenian Indigenous Religion" in this volume (xx).

[13] Rune Hjarnø Rasmussen, "Nordic Animism," YouTube channel, https://www.youtube.com/@NordicAnimism/videos

[14] For example, Ewa Domanska states: "…asking the questions 'what is an object?' and 'is it alive?' […] might support the emergence of an alternative worldview based on a participatory perspective of the world that understands (new) animism as a different way of being human and relating to the world on different terms." Ewa Domanska, "Is This Stone Alive? Prefiguring the Future Role of Archaeology," *Norwegian Archaeological Review* 51, no. 1–2 (2018): 22.

[15] "Like Creation stories everywhere, cosmologies are a source of identity and orientation to the world. They tell us who we are. We are inevitably shaped by them no matter how distant they may be from our consciousness." Krimmerer, *Braiding Sweetgrass.*

[16] Nordic Animism researcher Rune Hjarnø Rasmussen comments on the use of the Nordic mythologies and highlights that while the term indigenous should be reserved for marginalized societies, the old, pre-Christian, and animist mythologies can nevertheless be employed for future ethicalities, by rather using the term "traditional." Rune Hjarnø Rasmussen, "Is Nordic Animism Indigenous," YouTube, October 4, 2021, https://youtu.be/ivI9BxJyptg

While Latvian knowledges (as well as many other pre-Christian worldviews) were marginalized, the Latvian society has nevertheless undergone Christianization a long time ago and thus its mythologies can similarly be accounted for as "traditional" rather than "indigenous."

[17] Roman Zaroff, "Some Aspects of Pre-Christian Baltic Religion," in *New researches on the religion and mythology of the Pagan Slavs,* ed. Patrice Lajoye (Lisieux: Lingva, 2019), 183–219.

[18] Here and further, where I refer to Daina, unless specified otherwise, I will use the electronic Cabinet of Folksongs or Cabinet of Dainas at https://www.dainuskapis.lv. About the Cabinet of

including sexuality and relationship-themed folk lore, up to enchantments, cosmology, and other sacral topics. Since *Dainas* form part of the oral history of Latvia, the messages they contain have changed over time, and the original meanings are often lost or incomplete.[19] Nevertheless, they are very alive in the culture, and, on a personal note, have brought me strength and understanding of the world ever since I was a child, participating in folklore lessons as part of an educational group for 4–9-year-olds. *Dainas* are also still taught in school as part of language learning classes and cultural education, though their deeper meaning is often forgotten or left to self-interpretation.

When considering ethnographic material, one has to take into mind that *Dainas,* as well as other ethnographic materials, were encompassed many centuries after the Christianization.[20] This is, however, precisely the reason why, in my opinion, critical genealogy is the appropriate method of investigation, since its aim is not to recover or appropriate totally the past knowledges, but rather to illuminate the co-constitutive presence of alternate ontogenealogies and the usefulness of old epistemologies as for ethical futurity in *how*[21] to relate to the world.

The genealogies are thus always already unfinished and often cryptic and unreliable. Thus, I will partly rely on the work of ethnographers, with two interconnected aims. First, to employ existing analysis of the original meanings, where possible. And second, to foster interdisciplinary research of situated knowledges. Hence, in revisiting the past, I rely greatly on the research done by ethnographers and literature scientists that have dedicated their work to the research of old epistemologies. The most important sources here are Janīna Kursīte's books[22] and a recent course on the Latvian further learning platform "OpenMinded": "The Latvian World," as well as the articles by Sandis Laime, who explores nature themes in Latvian folklore. Thus, the research also employs the method of metanalysis, in reflecting upon ethnographic and folklore research in Latvia.[23]

Folksongs, see: Latviešu folkloras krātuve, "The Cabinet of Folksongs or Dainu Skapis," http://en.lfk.lv/collection/folksong-cabinet

[19] Zaroff, "Some Aspects of Pre-Christian Baltic Religion," 187–188.

[20] The Cabinet of Dainas (which is an actual cabinet constructed by Krišjānis Barons) which holds the largest collection of Latvian folksongs, was made in the late nineteenth century. While it leads some authors (such as Zaroff, 187) to conclude that ethnographic material is only usable as a supportive source (which might be true for the purposes of specific, fact–checking research), for a genealogical analysis this changing, transformative nature of genealogies is not hindering. See on Latvian folklore studies Sandis Laime and Beatrise Reidzāne, "Latvian Folklore Studies and Mythology", in *Latvia and Latvians: Collection of Scholarly Articles. Vol.2* (Riga: Latvian Academy of Sciences, 2018): 89–125.

[21] Rune Hjarnø Rasmussen refers to the myth as a "how" (with regard to relating to the world) rather than a "what" in "Ginnrunar and Netflix Ragnarok, A Review", *Nordic Animism* Youtube Channel, https://youtu.be/F5uyCBiokrw. See also: Yunkaporta, *Sand Talk.*

[22] Janīna Kursīte, *Latviešu dievības un gari* (Rīga: Rundas, 2020), Janīna Kursīte, *Latviešu folklora mītu spogulī* (Rīga: Zinātne, 1996).

[23] The article, however, does not propose to deliver an in-depth description of the meanings behind Latvian pagan traditions and knowledges but rather proposes rethinking old genealogies as a vehi-

Stone and forests bear significant sacral meaning in Latvian mythology. Stone is a significant reminder of the basis of becoming and life before societal configuration, while trees and especially sacral groves represent a way to organize time and space, as well as to distinguish between the living and the dead. I chose three instances for consideration—stone as the non-living beginning of existence, the idea of a "silver forest" in Latvian folksongs as a representative of sacral groves *not to be touched*, finishing with a reflection of dendromorphic transformation in pre-Christian thought.

Within this article, I focus on the elemental worldings of stone and forest, in the context of the agency of nature in Latvian pre-Christian worldings that express the affinity and affectual relations with (what is conceptualized as) nature still present in the Latvian society, considering old Latvian epistemologies as a fruitful ground for future affirmative ethics. In this road of investigation, I am led by the question: how does the presence of stones and forests co-constitute our understanding of life and nature?

Our Stone Beginnings

Latvian folklore has many references to stone as a significant reminder of the basis of becoming. Recurring is the theme of a stone in the middle of the sea, while water, stone, and forest (trees) are in a sense the three most sacral elements of old Latvian tales and lore, next to wildlife references to wolves, bears, and the like and relations with the sun, the moon, and the stars. Janīna Kursīte[24]

cites a Latvian Daina:God met devil
In the middle of the sea on a stone
Furs went crack, crack
Swords went clink clank.[25]

Kursīte's explanation of these lines, includes firstly the reference to the devil as the underground counterpart of God in the Latvian pagan worlding,[26] which later undergoes demonization with Latvian Christianization in the twentieth–fourteenth

cle for relating to the world today. Here, I follow the idea that myths are to be employed as a *how* to relate to the world, rather than accepted as a "what." Thus, I am only fragmentary mentioning some of the most necessary information on the epistemologies themselves, while focusing on their usefulness for today. While folk poetry itself has been researched quite broadly, interdisciplinary research that would connect contemporary philosophical thought with the situated knowledges of Latvian sacral scripture is lacking—and this is, thus, my attempt at trying to breach this disciplinary chasm.

[24] Kursīte, *Latviešu folklora mītu spogulī*, Ch. 1.

[25] Translated from: Sasatika Dievs ar velnu/Vidū jūras uz akmiņa/Kažociņi brikukraku/Zobentiņi šmigu šmagu (LD 33692) Here and further *Latvian Dainas—LD, cited from the Cabinet of Dainas* https://www.dainuskapis.lv/

[26] Kursīte, *Latviešu dievības un gari*, 231–234.

centuries. The fight between God and the devil in the middle of the sea (a mythical place of water and stone), thus, is rather the act of world creation than a fight between good and evil that underlines the significance of the chthonic world in cosmogony.[27]

As the place, where the devil (as the leader of the underworld) and God together create the world, the stone is also a place of sacral energy, of the energy of life before societal figuration. While it often, thus, also represents death, especially human death by a curse,[28] it is also a "prime matter."[29]

> As a place of sacral significance stone is a place/space where new sources of influence and energy can brew, as well as a vessel/precondition of condemnation, in cases of a breach of morality. In various sources, the goddess Māra, in the form of the grass snake is rolling upon a stone. In a folksong, as she rolls, so will the soul of any man who humiliates girls:Black snake rolls around
> Middle sea on a stone:
> As will roll the soul of the boy,
> Who disdains a maiden[30]
>
> With this, stone and soul are tightly interlinked—rather than antonyms the non-living and the living are one, and the agency of the non-living is to be taken into account and respected. Yet another Daina says:The black snake draws a thread,
> Through a grey stone.
> I drew out a fellow's mind,
> Sleeping with, for one night[31]

The song suggests that the maiden has "pulled out" the mind of a fellow after sleeping with him once. This Daina seems to add at least one other meaning

> to the significance of stone—namely, as a symbol of destiny. The most famous modern adaptation of these Daina's is performed by the folk-rock music group "Iļģi" which puts together both Dainas, adding another one:The black snake draws a thread.
> Over a grey stone,
> I pulled out a fellow's mind,
> Sleeping with for one night
>
> I wrapped in the fellow's mind
> In a white clue,
> So that he could not

---

[27] See also more on the chthonic world in Latvian tales Ingus Barovskis, *Htoniskā pasaule latviešu pasakās: laiks un telpa.* [*Chthonic World in Latvian Tales: Time and Space]*, Doctoral dissertation, (Rīga: University of Latvia, 2015).

[28] Barovskis, *Htoniskā pasaule latviešu pasakās,* 70–72.

[29] Barovskis, 183.

[30] Translated from: Melna čūska riņķi grieza/Vidū jūras uz akmiņa/Tā griež puiša dvēselīt'/Kas meitiņas nicināja. LD 8694.

[31] Translated from: Melna čūska diegu vilka/Caur peleku akmentiņu/Es izvilku puišam prātu/Vienu nakti guledama. LD 10296.

Pull me every which way.[32]

Black adder rolls around,
Middle sea on a stone:
As will roll the soul of the boy
Who disdains a maiden"[33]

Seen together like this, the three Dainas showcase relations between stone, life, judgment, femininity, and sacrality, in particular, within the figure of the deity Māra.[34] Māra is the ancient earth goddess in Latvia—known for bestowing as well as taking life. She is often seen in the form of a black snake, namely a grass snake (like in the folksongs above), but she can also be seen as a black cow, and in other forms, and prayed to in different forms, including as the Mother of all creation in the form of a stone.[35] Moreover, the existence of the popular adaptation inspires the employment of Dainas in current worldings as well as the persistence of ancient genealogies, including in the context of feminine strength.

As these Dainas show, the tone in Latvian pre-Christian thought represents the first beginning, the static, unchanging essence,[36] the beginning of every land piece, as well as the cosmos itself, embodying not only the non-living static landscape but also the potentiality of life within the non-living. Sacral stones are still present in the landscape of Latvia.[37]

---

[32] This addition seems to be transformed from a different Daina: LD 9646, which comments on master-serf relations with "I wrapped the master's mind so that he could not give me to a different master."

[33] Melna čūska diegu vilka/Pār pelēku akmentiņu/Es izvilku puisim prātu/Vienu nakti klāt gulēj'si.
Es ietinu puiša prātu/Baltā diegu kamolā/Lai tas mani nevarēja Pa prātami dancināt.
Melna čūska riņķi grieza/Vidū jūras uz akmeņa/Tā griež puiša dvēselīte/Kas meitiņas nicināja
See, Iļģi, "Vissbija," Album "Saules meita" [*Daughter of the Sun*], Youtube, https://youtu.be/O32ZjGfcy4A

[34] Kursīte, *Latviešu dievības un gari,* 127–137. Interestingly, the Latvian pre-Christian religion excels within the pagan religions of the nearest Prussian and Lithuanian regions with the smallest number of male deities.

[35] Janīna Kursīte. Lecture 1 "Latvieša pasaules uztvere un ticība" [Latvian worldview and beliefs], *Latvian World,* Lecture course, OpenMinded, Riga, Latvia).

[36] Barovskis, 70–72.

[37] It is also of note that the post-soviet period has seen a proliferation of "new-found" sacral places in the Latvian landscape, i.e., the so-called "modern holy places". These are criticized by folklorists, scientists, and other skeptics as merely a "good business plan" and are thus problematic in the least yet remain popular. Rūta Muktupāvela, "The Mythology of Ethnic Identity and the Establishing of Modern Holy Places in Post-Soviet Latvia," *The Pomegranate* 14, no. 1 (2012): 69–90. While these popular places remain at least controversial, it is yet a possible indication of the necessity of a new animism or a new relation to the earth felt by the people.

Are the Stones *Still* Alive?

Agency of the stone today represents the discussion of life and non-living as a problematic dichotomy,[38] especially in the context of death as an instance in life, demonstrated as its opposite only in the context of a self, conceived in the context of a human-centered substance ontology.[39] While life and death already lose their oppositional value via critical posthumanist worldview of processuality and continuity beyond the individual,[40] non-living and life still represent a dichotomy that is to be rethought to realize the entanglement of becoming.

The sacrality of stone in Latvia and the nearest regions (where stone "baba"[41] represents an old stone personae of ambiguous gender) allows reconsidering the idea of stone and nature-beyond-life as senseful and agentic forces co-constituting human and more-than-human life. Moreover, the sacrality of stone also symbolically denotes the value of naturecultures beyond and before I-conscious decision, suggesting the meaning of simplicity, as referring to more knowledge than words.[42]

The stone refers to essence as slow time that is nevertheless in process and transformation. As such, it stands in opposition to plastic as an object *par excellence,* a cancerous fixedness of matter in the non-destructible (life-everlasting),[43] and represents the potentiality of life within non-life, but also allows understanding of naturecultures beyond the dialectic of civilization and wilderness,[44] as a tool-on-the-fringe of life (also quite literally, in the context of the Stone age).

---

[38] On life and non-life: Radomska, *Uncontainable Life*. In context with geological becomings see Elizabeth Povinelli, *Geontologies: A Requiem to Late Liberalism* (Durham: Duke University Press, 2016).

[39] Anne Sauka, "Selfhood in Question: The Ontogenealogies of Bear Encounters" Open Philosophy 5, no. 1 (2022): 532–550.

[40] And rather reveals a dance of the Apollonian and Dionysian beyond the dialectic, Gilles Deleuze, *Nietzsche and Philosophy* (London: Continuum, 1983), 10–17. My work against the dichotomy of life and death see Sauka, *Life in Process,* Anne Sauka "Becoming Self: A Legion of Life in a Culture of Alienation." In *Normalitāte un ārkārtējība filosofiskā skatījumā* (Rīga: LU Akadēmiskais apgāds, 2022), 25–46.

[41] Janīna Kursīte, "Latvieša pasaules uztvere un ticība."

[42] Going even further, this idea then could be traced back to *Leib* phenomenology or Nietzsche's idea of the lived body as the big mind that is more knowledgeable than the consciousness that reflects upon it (as consciousness is just one of its expressions, i.e. the "small mind"). Nietzsche, KSA, 4:39.

[43] At least in a symbolic context. On the ontology of trash as a non-existent that persists and as such is to be contrasted to the existent that does not persist see: Greg Kennedy, *The Ontology of Trash* (Albany: State University of New York Press, 2007), x.

[44] See Ramachandra Guha's critique of (local/American) wilderness/civilization dichotomy and its harmfulness for global contexts: Ramachandra Guha, "Radical American Environmentalism and Wilderness Preservation: A Third World Critique." *In The Future of Nature. Documents of Global Change*, edited by L. Robin, S. S Sörlin, P. Warde, (New Haven: Yale University Press, 2013), 409–426. See also William Cronon, "The Trouble with Wilderness; or, Getting Back to the Wrong Nature." *Environmental History* 1 no. 1 (January 1996): 7–28—a discussion on the changing understandings of wilderness in the Global Northern/Western contexts.

The stone, thus, could be termed as -jectness enfolded—lifeless stillness that secures potentiality. I, here, use the concept of "-jectness" as it is employed by Michael Marder in "Philosophy of Dump"[45] where it refers to a misguided materialism that oversteps the dichotomies not by a differentiation but via unification, thus, purging subject and object from their meaning and replacing that with a non-meaning, a sort of -jectness—the materialization of which might be the amorphous mass of dump/landfill, i.e. also the -jects par excellence in dumped plastics that become trash with the moment of their conception. "-jectness enfolded," thus, refers to the opposition of -jectness in potentiality, the not-yet-agency of a stone that can become-tool, a space of happening, or become-life or a place of belonging, and thus, is truly a -ject.

The story of the fellow and his tortured soul for his despise of the girl, brings another layer to the hitherto contextualized old epistemologies, strengthening the link between environmental humanities to feminist philosophies. While Latvian epistemologies deliver a strong focus on female goddesses,[46][47] thus, opening them for feminist discussion, especially in regard to human-nature embeddedness[48] the presence of the various "mothers"[49] that claim the space of the sacral can also be

[45] Michael Marder, *Dump philosophy* (London: Bloomsbury, 2021), 4–5.

[46] An important discussion here would be to reframe the hierarchy of religions. Most often, the "development" of a religion is judged based on the question, of whether the religion has a developed pantheon. In essence, this is a presupposition that all non-monotheistic religions are hierarchically lower than Christianity and other monotheist religions. This view is also expressed by Leons Gabriels Taivāns, "Baltu reliģija un kristietība (tipoloģiskas rekonstrukcijas principi)" [*The Baltic Religion and Christianity (Principles of Typological Reconstruction)],* Ceļš no. 44 (1992), 18–32, where he discussed the "lacking" of morality and understanding of the "state" in pagan religion. Similarly to how James Scott, *Against the Grain: A Deep History of the Earliest States* (New Haven: Yale University Press, 2017) highlights how the derision and fall of early states might be only considered as a loss from the point of view of states, not the point of view of the people, more "dispersed," *Dionysian* religions might also represent a less controlled society that has the same faith, while different practices, different choice of main deities, etc. It could even be argued that a developed pantheon goes hand-in-hand with patriarchal religions. This is certainly true for Latvian vs Lithuanian religious practices, where Lithuanians have a more developed pantheon with male deities having a stronger role, and the opposite is true for Latvian mythology. Notably, Lithuanians accepted Christianity on their own accord for political gain (twice), while Latvians were more or less forcibly Christianized. Zaroff, "Some Aspects of Pre-Christian Baltic Religion," 186.

[47] See also the work of archeologist and anthropologist Marija Gimbutas, who has written extensively on goddesses and the role of femininity in religion/culture, as well as on the Baltic pre-Christian religion. On the Goddess Laima (Luck/Happiness) see Marija Gimbutas, *The Lanugage of the Goddess* (San Franscisco: Harper & Row, 1989), 111. More in Marija Gimbutas *The Balts*. London, 1983; more on her work Joan Marler, "Baltic Archaeology, Cultural History, Ancient Lithuanian Symbolism, Old Europe, and the Archaeomythology of Marija Gimbutas," *Archaeologia Lituana* 23 (2022): 10–33.

[48] Carolyn Merchant, *The Death of Nature. Women, Ecology and the Scientific Revolution* (San Francisco: Harper & Row, 1990).

[49] Kursīte, *Latviešu dievības un gari,* 142–166.

viewed critically as an essentialization of womanhood, leading to a different, yet *the same*[50] binary, where nature is equated with women.[51]

Yet, alternate ontogenealogies (especially with regard to seeking ethicalities for the future) often open horizons for reworking and rethinking the question of gender in context with environmental practices, attitudes, and ontologies, making, if not all, then a large part of environmental humanities feminist.[52] That is not to glorify old epistemologies for their feminist standpoints—rather, although they might have some very common traits, their ontologies also broaden the perspective of ontological reflection and ways in which and *how* to relate and participate within the world, thus, already weakening the univocal ratio-centric[53] homogenization of the world, which, in itself, is a feminist philosophical move. From this standpoint, the story itself is, thus, another example of how feminist and environmental topics are closely related: (1) with regard to their common marginalization, (2) with regard to the devastating consequences of their marginalization, (3) with regard to the heterogenization of knowledges that is instigated by the remembrance of old epistemologies.

However, here I would like to shift focus towards a more concrete subject, namely, the role of the stone in the story about the fellow and the girl, to demonstrate not only the overall relevance of feminist themes and topics to environmental humanism, which has scrupulously been done before,[54] but to also highlight some of the aspects that relate to the affectual level of the story itself and the ontogenealogies we live by.

Here, I think, the feminist themes come more concretely in focus, while simultaneously tying in with the environmental considerations of relatedness to embeddedness, and the *slow time* of stone, when considering the goddess's stone as a symbol of justice. Stone as a *beforeandafter* is both the place of creation as well as the place of remembrance. It is, thus, a special place of justice for the unjustified—an impossibility enacted. Drawing a thread *through* or *from*

> the stone appears as a mythological imaginary of the impossible, as well as a cautionary tale against a binary A is not not-A logic.Another very popular folksong states:
> On the stone, I cut the firewood,
> I lit the fire in the stream,
> Let there warm up those folks

[50] Using the term in an Irigaraian sense. Luce Irigaray, *This Sex Which Is Not One*, trans. Catherine Porter (Ithaca: Cornwell University Press, 1985), 74.

[51] See Alaimo, *Bodily Natures*.

[52] Lauren LaFauci and Cecilia Åsberg, "Is All Environmental Humanities Feminist Environmental Humanities?" Seeing the Woods: A Blog of the Rachel Carson Center for Environment and Society (July 6, 2020). https://seeingthewoods.org/2020/07/06/is-all-environmental-humanities-feminist-environmentalhumanities/

[53] That is not to say that "rationality is bad," but rather to highlight the necessity of plurality and perspectivism in thinking.

[54] LaFauci, Åsberg, "Is All Environmental Humanities Feminist Environmental Humanities?"; Greta Gaard, "Indigenous Women, Feminism, and the Environmental Humanities," *Resilience* 1 no. 3 (Fall 2014): 86–99.

Those, that do not wish me good[55]

"Those folks" are sometimes replaced by "mother-in-law" (of either side of the marriage), while the action is sometimes done by a black or colorful adder grinding flour on a stone.[56] This "wish" delivered a very strong message, perhaps used also as a dark spell,[57] and should not be taken lightly, while also demonstrating the possibility of the impossible. "Fringes" in Latvian epistemologies are places only for the wise, while it is safer (and more ordinary) within the middle of the space.[58] Perhaps, here also, the possibility of the impossible indicates a point of margin, a fringe—the place for the philosophical and the sacral, not to be taken lightheartedly.

The stone as a place for the impossible, while also a place for justice, thus, allows thinking of justice in remembrance. When there is no justice, is there justice after all in precisely its non-being, its silent, bracketed -jectness? A lack as a memory? This interpretation ties in also with the morphisms in stone that are often part of mythologies—thus, in 1965 the Latvian landscape lost its most famous (and only real) waterfall—Staburadze due to the building of a Hydro station. This waterfall, similar to other cliffs in the world, also bears an old tale of a girl, whose fiancé—a fisherman dies in the river fishing, after which she stays there on the riverbank, and turns into a stone, but the tears do not stop flowing. This is, most probably, a newer, already romanticist tale, nevertheless its message, seems to mirror the idea of stone as not only a place of potentiality or creation but also a place of remembrance.

Thus, while alternate stone-human relationships today are mostly concealed by the dominating dialectics of wilderness and civilization in the Global North that depicts non-life as belonging to human-created environments, with lifeless housing and cities, the genealogies of stone beginnings to becoming continue co-constituting materialities and could allow alternate ethicalities for future posthuman reflection of city-wilderness co-creations.

As I tried to show in my reflections on the conceptual context of stone becomings, the stone is a marker of the possibility of the impossible and the *beforeandafter* life, and as such:

---

[55] Translated from: Uz akmeņa malku cirtu/Strautā kūru uguntiņu/Lai sildās tie ļautiņi/Kas man laba nevēlēja. LD 9121–2.

[56] Barovskis, 72; Jolanta Stauga also highlights that with the act of grinding and rolling something gets remade: "…cosmos is distinguished from the primal chaos. Thus, the sake can be looked upon as an ambivalent being, where, on the one hand, it is the one that creates a cosmogonical space in the middle of the primal waters, while, on the other hand, the snake is associated with chaos and can create it also itself, which is signified by its black color and its belonging to the chthonic world." Jolanta Stauga, Dzīvnieki latviešu folklorā: Mitoloģiski maģiskais aspekts (pēc latviešu tautasdziesmu un mūsdienu folkloras materiāliem) [*Animals in Latvian Folklore: Mythological Magical Aspect (After the Materials of Latvian Folksongs and Modern Folklore*], Doctoral dissertation, (Riga: LU, 2011), 85.

[57] Kursīte, *Latviešu kultūra mītu spogulī*, Ch. 3

[58] Janīna Kursīte, *Mītiskais folklorā, literatūrā, mākslā [The Mythical in Folklore, Literature, Art*] (Rīga: Zinātne, 1999), Chapter: "Mītiskā telpa" [Mythical Space].

(a) *stands* as a marker of naturecultures—where it signifies the possibility of tools, thus, demonstrating the necessity of viewing technology beyond the dichotomy of wilderness and civilization, while also being firmly a part of nature;
(b) *resounds* the sacrality/sensefulness of naturecultures—where it signifies the *slow time* of remembrance in the non-memory of the *after*;
(c) *resonates* creation, as a symbol and materialization of potentiality in the non-time of *before.*

Thus, remembering the stone as a potent marker of the naturecultured aspects of human and more-than-human becoming can be both experientially as well as philosophically fruitful.

Silver Forests *Not to Be Touched*

Next to stone, forests also bear sacral meaning in Latvian mythology, with sacred groves[59] and trees representing a way to organize time and space, as well as to distinguish between the living and the dead.[60] The Latvian equivalent of the "world tree" is Austras koks which symbolizes the Sun's path and the order of the world, with its roots digging into the underworld, leaves growing into the sky world, and the trunk representing the *middle,*

> where humans are living. The figure of the sacred tree is also represented in folksongs. A folk song goes:The sacred tree
> Has nine branches
> On the top of each branch
> Nine flowers.
> On every flower
> Nine berries.[61]

In the context of ancient Latvian understandings of time and space,[62] Janīna Kursīte explains that Nine x Nine x Nine—represents 729, which, considering that day and night in old Latvian culture were counted separately, makes up a year. The

---

[59] Sandis Laime "The Sacred Groves of the Curonian Ķoniņi: Past and Present," *Folklore (Estonia)* 42 (2009): 67–80. Ķoniņi was a unique group of Latvians living in the Courland region, who never experienced serfdom and in this sense were, in fact, Latvian "landlords." In effect, they were the last who had sacred groves until Soviet occupation, and thus their practices have attracted research attention. "The Curonian ķoniņi were neither peasants nor free peasants, but instead constituted a special estate of freemen, to whom the peasant laws of Courland did not apply and who enjoyed the same status as other vassals of the order in Courland." (Laime, *The Sacred Groves,* 67.)

[60] Janīna Kursīte. Lecture 2 "Latvieša laika un vietas uztvere. Svētbirzis jeb "dots devējam atdodas"" [*Latvian understanding of time and space. The sacred grove or "you get what you give"*], *Latvian World,* Lecture course, OpenMinded, Riga, Latvia.

[61] Translated from: "*Svētajam kokam/Deviņi zari/Ik zara gala/Deviņi ziedi/Ik zieda galā/Deviņas ogas.* LD 33601.

[62] Kursīte, *Latviešu folklora mīta spogulī,* Ch. 3.

combination *trejdeviņi* (literal translation: three-nines) gets repeated and repeated in ethnographic materials. It is, for example, well known that when preparing for the summer solstice women and girls are supposed to make their flower crowns from 3 × 9 different flowers[63] and weeds (as every plant is "good" on the Jāni's festivities eve).

Moreover, sacred groves and sacred trees also organize time and space, with sacred tree sometimes designating the middle.[64] Thus, the tree in the folksong encapsulates both time and space, and rightfully so, since while they live long, trees are both the guardians of space, as well as representatives of the changes of time within their growing processes. Thus, if water and stone get mentioned as "everlasting"—literally "living the lifespan of the Sun," and are thus contrasted in a popular folksong to the mortality of the human being,[65] then trees speak to the transformative aspects of life (and the cosmos in a broader sense), and as organizers of time and space stand on the margins of sacral and secular space. Moreover, where stone represents the initiation, the beginning of time and space, trees are the extension—the organizing feature of the time-space.

As mentioned earlier the middle is usually the safest place, while the fringe is something uncanny yet—it can also be sacred (or philosophical, perhaps?). The sacred grove is, thus, also associated with what is crooked—it is not something skewed or wrong, but rather a way around, as opposed to everyday dealings that just go straightforward, which is why sacred groves get the definition of being *elka* groves, associated with a "turn."[66] The sacred is the philosophical, the senseful materiality taking part in the symbolic co-constitution of the world.

Trees are thus the sacral organizers of time-space in Latvian mythology—they distinguish between the living and the dead, the sacral and the secular (with *elka* or sacral groves), the safe middle and the philosophical margins on the threshold of the sacral world beyond life. This role of trees is reflected also practically in some of the funeral rites that include marking special trees that one should not touch and that draw the "line in the sand" between the living and the dead so that the living would respect this boundary and keep on living rather than revisiting, again and again, the past.[67]

[63] Usually, today that is interpreted today simply as "many" different plants, and each year I promise myself to count, which I do not…

[64] Kursīte, "Latvieša laika un vietas uztvere."

[65] Dzīvoš ilgi, ne tik ilgi/Saules mūžu nedzīvošs/Ūdentiņš, akmentiņš/tie dzīvoja Saules mūžu. (*Live I long, or not that long/I won't live the life of Sun/Water and stone/they will live the life of Sun.* LD 27341-2.

[66] "Elka: is also translated as an idol, which is why idol worship in Latvian is "elkdievība" or the wrong, pagan faith, according to Christianity. "Elka" is also associated with the Latvian word for "elbow"—*elkonis,* and thus, with a "turn," with something that is not straightforward, but rather a "way-around", such as symbolical interpretation or divination. Janīna Kursīte "Par Austrumprūsijas un Kurzemes vietvārdu stāstiem" [On Eastern Prussian an Courland's Toponyms] *Latvijas Vēstnesis* No. 183, 2003.

[67] Kursīte, "Latvieša laika un vietas uztvere."

Where Did All the Trees Go?

When thinking of the meaning of nature-human relationships, trees also seem to signify the border at the edge of the entanglement of nature, culture, and technology—a sort of signpost that grows through these dimensions that nevertheless layer on top of each other, thus, representing a united whole. Here, I refer to a folksong that stands out for me and showcases a tree as the epitome of the multidimensionality of embodied experience:

> I went through a silver grove,
> Did not touch a single branch,
> If I would have broken one
> I would walk in silver.[68]

Although the folksong probably refers to a sacred grove, where multiple sources warn from breaking branches,[69] it seems that this folksong does not have a single right interpretation, since it starts an important discussion, where unnecessarily touching a branch, does not end in misery or a problematic situation, but rather affords one the status of walking "in silver."

This goes beyond the dialectic of reward and punishment, when deciding upon changing the natural flow of things, and highlights an alternate significance of not touching what is not yours. Thus, "walking in silver" might represent the gains that are today associated with the deterioration of nature as a resource in today's affluent societies and denote the usefulness of trees for humans. Yet, where "walking in silver" might seem a positive outcome for today's societies, the parable of the folksong probably goes further than obvious rights and wrongs, denoting both the resourcefulness of nature, as well as the necessity to preserve.

A group of other folksongs also represent the relations between humans and nature, where the main motive is a tree talking to a girl and inviting her to take all the branches she needs while cautioning and pleading with her to not take the top of the tree.[70] Other songs talk about a tree and a fellow or a girl arguing with a tree because of the breaking of the top of the tree,[71] and in yet others, the young herders are perplexed, as to why the tree grove is crying, if they have only taken a few branches for the flower crown.[72] While these are less mysterious as to their

---

[68] Caur sidraba birzi gāju/Ne zariņa nenolauzu/Būt zariņu nolauzusi/Tad staigātu sidrabā. LD 30637.

[69] Laime, "The Sacred Groves of the Curonian Ķoniņi," 70.

[70] Songs about the top of the tree: Bērziņš mani aicināja/Nāc, meitiņa, slotas griezt/Griez, meitiņa, zarus manus/Galotniti vien negriez/Lai palika galotnīte/Vējiņam vēdinàt. [*Birch tree invited me/come and cut my branches, girl/just don't touch the top/allow the top to stay/for the wind to blow through*].

[71] Es izdzirdu ganu puisi/Ar bērziņu baramos/Bērziņš saka: zarus griež!/Gans nogrieze virsuniti. [*I heard the herder boy/Quarrelling with the Birch tree/Birch tree says: branches are cut!/Herder cut the top.*] LD 2767-1.

[72] Kalabad tu, bierztaļa/Uz manim žēli raud'? /Nosalauzu vienu rīksti/Svētu rītu ganīdama [*Why do you, grove/Cry to me in sorrow?/I cut only one switch/herding in the holy morning.*] LD 2762-5.

meanings, they mirror the message of "walking in silver" as well as reflect the lively relations with trees.

Here, while the recognition of individual gain is present, it does not dominate other kinds of gains, allowing a double interpretation of

1. both recognizing the gifts of nature in the creations of humanity and
2. reflecting their necessity and usefulness with respect to the agency of non-human beings—in this case, the trees.

Thus, this folksong refers to an ever-present affectual experience of being in awe of nature, while simultaneously foregoing it by "outsmarting" natural flows as a singular non-contradictory experience and recognizes the entanglement and dependence of humanity upon already present more-than-human others, while simultaneously also recognizing human agency and responsibility.

Today, the processuality of wood products, is significant and still present in many Baltic and Eastern European countries, especially in house heating and building contexts, and does constitute the basis of this folksong, where choosing one or the other is not a straightforward decision, but rather a reflected move that includes both touching and remaining in close entanglement, and the notion of rather leaving untouched, where possible. This ethical maneuver goes beyond either/or to the territory of beyond dialectics and could be helpful for a posthuman understanding of human-environment relations.

Moreover, in Latvia relations with trees, today are strained between two opposite yet seemingly similarly alienated discursive outlooks. On the one hand, "Latvian State Forest" propagates the idea of the forest as a "bed of carrots" that need human tending and harvesting, while, on the other hand, (mostly urban) "forest lovers" focus on the fragility of nature (thus, also stripping it of agency)[73] and saving "the wilderness." Both opposites thus uphold a dialectic of wilderness[74] and civilization, while retaining human exceptionalism. These have various genealogical roots, from the agrarian romanticism that equals nature with a garden for human tending, to more industrial discourses stemming both from Western industrialism as well as Soviet proletarianization of nature.[75] Undeniably, also the past sacralization of trees in a transformed way today might have an impact on the imaginaries that equate nature mostly with plant life, and especially the forest, thus, building imaginary boundaries between the planet *per se,* nature as the living realm beyond humanity and civilization, and the human being. I remember hopping off of the plane in Iceland and having the immediate reaction—where did all the nature go? Obviously, making the association between trees and nature. Still today, to be "at home" means

---

[73] Katherina Schwartz, "The Occupation of Beauty: Imagining Nature and Nation in Latvia," *East European Politics & Societies* 21 no. 2 (2007): 259–293.

[74] The problematization of wilderness—Guha, "Radical American Environmentalism," Cronon, "The Trouble with Wilderness." I discuss the dialectic as a *status quo* in Sauka, "Selfhood in Question."

[75] Anne Sauka, "Breaching the Dialectic with Situated Knowledges: The Case of Postsocialist Naturecultures," *Polish Journal of Aesthetics,* 2023.

being within the safe harbor of trees, as many Latvians pride themselves on their mushrooming and berry-picking practices, and the forests are free for roaming to anyone, beyond the factual property rights.[76]

The association of plants as the factual representatives of nature is also reflected in the traditions of gardening that showcase Baltic nature imaginaries as entangled with a strong "garden culture," with the almost obligatory flower garden by every house.[77] The delineation between plants as nature vs planetary and human realms, though, has proven intangible to relate the environment beyond the limits of human/environment differentiation, thus, the older sacralization of trees, as well as the relations between the planter and the soil might prove a fruitful ground for reconsideration of the transformed genealogies, and seeing the trees as constitutors of our livelihoods, entangled with human practices.

Yet, a consideration of the positive impact of the association between trees and nature is also a significant discussion point. When thinking of the agency of nature itself and the fact that more than half of Latvia's territory is clad in forest,[78] it is clear that the environmental surroundings co-constitute this immediate affection and association. As the organizers of time and space, trees are, thus, not only one of the main symbols of what is most easily associated with "nature" in Latvia but also importantly breach the dimensions of nature and culture. Hence, the practices that are already in place (such as gathering, gardening, and composting[79]) that today mostly associate with the dialectic of (a) the agrarian romanticist (itself a dialectic of sorts) or (b) industrialist attitudes and genealogies, might regain their sacral dimension in organizing human space and time within natureculture. Rather than as a silent, aesthetic spectator in awe or a resource extractor, humanity is and should be thinking and becoming *with* the forest, through inherited practices.

Another important factor, when considering the organization of time and space via trees, is the mirroring that some folksongs set in place, talking about humans as

[76] Similar to the rest of Europe but in stark contrast to the United States of America, where (1) the forest might be someone's untouchable property or (2) the forest might be the "property of nature" in the context of wilderness conservation.

[77] Especially in the past, yet this tradition is still quite strong for private housing that rarely has an empty lawn, and mostly has a smaller flower garden beside the house, as well as at least some fruit trees, vegetable patches, and berry bushes behind the house further inside the garden.

[78] LSM.lv "Mežu platība Latvijā turpina pieaugt—pērn tie bija 52%" [*The Area of Forest keeps on Increasing Last Year it was 52%*]. https://www.lsm.lv/raksts/zinas/latvija/meza-platiba-latvija-turpina-pieaugt%2D%2Dpern-tie-bija-52.a325062

[79] Petr Jehlička, Miķelis Grīviņš, Oane Visser, Bálint Balázs (2020), "Thinking food like an East European: A critical reflection on the framing of food systems," *Journal of Rural Studies* 76 (2020): 286–295.

trees—dendromorphizing the human experience[80] or making similes between the tree and the human world.[81] Thus, these songs both:

1. place the human "within the trees", i.e., showcase humanity as part of the natural world, and symbolically—as part of the "trunk" of the *Austras koks,* or the world tree— spacetime dimension that transgresses the boundaries of the various dimensions in terms of disposition, i.e., similar, to how we could imagine that the designation of something as "one" simultaneously assumes the existence of zero and two, as well as
2. delineate human belonging to the living, mortal world, as part of nature, growth, and transformation of becoming, more similar to trees that, for example, stone or water— spacetime dimension that describes and prescribes the becoming human, similar to how we could imagine describing the properties of "one."

As I exemplified in my considerations on the conceptual context of tree becomings, trees are the marker of the human-nature relationship, and the process of life in its presence, both organizing and delineating time and space (and thus placing humanity in a particular yet transformative place within cosmos) as well as transgressing the dimensions. Thus, while stone stands for the *beforeandafter* and the slow time of the Sun, trees denote the (sacral) interconnectedness of various planes of becoming, while also organizing this multidimensional space and "clearing" the space for human activities and livelihoods. As such, the tree:

(a) *reflects* the relations with nature—where trees stand as a *between* of the different realms, while also asking for caution in interaction, thus, securing differences between humans and trees, while also delineating their entanglement.
(b) *mirrors* the human being as part of the living world of growth and decay, transformation, and rites of life (where trees factually also are chosen as the signifiers of different rites of life[82]).
(c) *organizes* time and space and designates human place within it, while also retaining the wholeness of the different planes (underworld, sky world, living world as well as nature, culture, and technology), demonstrating that present

[80] Izstaigāju bērzu birzi/Lāga rīkstes nenogriežu/Izstaigaju Bebrenješus/Lāga puišu neredzeju/Cits bij greizs, cits bij šķībs/Cits meitiņu salauzits. [*I walked through the birch grove/Did not cut a [proper] switch/I walked through Bebernieši/Did not see a proper fellow*] LD 12756-1.

[81] Precību dziesma: Tumši, tumši tie mežiņi/Kà nav tumši, egles vien/Sveši, sveši tie ļautiņi/Kà nav sveši, tautas vien. [*Dark, dark those forests/how could they not be—only fir trees/Foreign, foreign those people/How could they not be—only folks*]. LD 18910-2 This folksong does not translate that well and needs further explanation. Namely, it could mention that there is not a single oak tree "to provide light," while this remains omitted in the particular verse. Also, "only folks" here means anyone outside of the immediate circle/relatives of the girl that is singing the song, thus, basically, it is a song about being married away, since "the coming of folks" usually denotes the coming of eligible bachelors, and "giving to folks" means marriage.

[82] Not only death but also birth. The goddess Māra is said to have a chamber where she acts as a *primum mobile* or a first mover for (wooden cribs). Kursīte, *Latviešu dievības un gari,* 129. Pilla Māras istabiņa/Sīku mazu šūpulišu/Vienu Māra kustinaja/Visi līdzi līgojàs. [*Full is Mara's room/ with small cribs/She moves one/All move together with it*] LD 1872, 1.

action, while situated in space and time, is at the same time also a becoming, a happening transgressing the different planes.

Thus, trees as a potent marker of the naturecultured aspects of human and more-than-human becoming can be vehicles to *think with* philosophically, as well as partake in the affectual relations of humans and their environments.

A Final Crossing: Dendro-Techno-Morphic Transformation

A popular folksong in Latvia tells a tale of the youngest daughter of a mother drowning in a river, being thrown out on the riverbank, and becoming a Linden tree.[83] The tree is then cut, and Latvian traditional zithers (Latvian: "kokle") are fleshed out of it. The *kokle* sings so poignantly and beautifully that the mother recognizes her lost daughter singing. On the surface, this song is probably one of the many descriptions of initiation into adulthood, since folksongs referring to either the initiate (the youngest daughter) or their parent dying are usually of this nature and do not imply death in its literal sense. The Linden tree is thus a totem that designates growing up and getting ready for marriage.[84]

Nevertheless, this song with its many variants allows philosophical reflection of the imagery employed to describe initiation, and as such is one of the remnants of a long lost "pagan" worlding that tells a curious tale of dendromorphism continued in technomorphism and speaks of the agency of the musical instrument. The object enchantment can be critically interpreted as an anthropomorphism of technological tools yet can also serve as a genealogical marker of sense-making of processual materialities beyond an anthropocentric worldview.

In this context, the tale as well as the initiation itself refers to life beyond death in technique,[85] beyond nature-technique dialectic, reflecting upon human-created worldings as part of the naturecultured lifeworld, within senseful and spirited as well as material context. Moreover, the techno-dendro-morphic transformations retain the soul of the daughter within the singing—a crucial part of Latvian culture, still retained today and reflected in the modernity, for example, in the "Latvian Song and Dance Festival" as well as the Singing Revolution in the Baltics. Here, life remains, while the forms wither away, change, and transform, not as a form of wilderness, but as part of human/more-than-human lifeworlds.

In the long-lost lore of the past, the link between (human) embodiment and the environment and thus also the understanding and experience of life and nature is

[83] See LD 33625-9.

[84] Janīna Kursīte, *Dainu kodekss* (Rīga: Rundas), 369–370; on the meanings of *kokle*: 203–204.

[85] Although the "death" here is symbolic, it is peculiar how the *kokle* retains the agency of the youngest daughter, especially for the mother. Is it a symbol of what her daughter left behind or is it her now married daughter that she only "hears" and does not see? In any case, the agency of the non-living comes to the fore.

radically different from the common assumptions, yet they are a significant ontogenealogical marker for an environed embodiment that is still lived and breathed in every second we live. Humans, here, as well as human creations, are part of the lifeworld, enablers of the agency of technology, as well as subjects to this agency beyond any dialectic of subjection and objectification.

Thus, these past epistemologies are a possible source of futurities, when accompanied by embodied critical thinking or an alternate sense of body-environments that allows a peek in the beyond of habitual lived conceptualizations and materialities.

Today, when human-stone and human-forest relations are largely de-enchanted,[86] touching a stone or the bark of a tree might not achieve the effect of the same affectual relatedness to non-living. Moreover, the ideas of "faith" and "virtue" today have traveled far from their immanent and non-abstracted meaning in pre-Christian thought, where, at least for ancient Latvians they were thought to be linked with the idea of usefulness.[87] Thus, while there is a de-enchantment of the lifeworld around us, alienation from what it means to be faithful (to Earth, i.e., immanently) is also present, making it a twofold predicament of loss of affection and enchantment resulting in alienation. Where enchantment was related to practicality and application, disenchantment has resulted in alienation and transcendental trickery. "Faith" becoming uprooted from its practical yet enchanted materiality, loses grip and becomes an empty promise, while the materiality also loses its sensefulness and becomes empty rigidness. This connection of both materiality and faith, when untangled, leaves both to become empty concepts without their senseful material grip. A re-enchantment or re-storying[88] of materiality could therefore bring faith back to its material rootedness, while simultaneously bringing the spirit back to materiality, going beyond the dialectics of transcendence and materiality.

The story of the girl-becoming-musical instrument recognizes faith as a lived experience, embodied in the materialities-we-live-by, rather than disconnected or discriminating against what kinds of human-created or non-created materialities the sensefulness of spirit can be attributed to. This reflection of spirit or vibrancy of matter[89] in human-created technology also reflects today's worldviews of the law of conservation of energy, and anchors faith again within the embodied experience.

The popularity of this song demonstrates how the present ontogenealogies are co-constituted by the past lore, showing human-non-human-more-than-human relatedness as an embodied and fully respectable link, beyond sacrality as no-saying

---

[86] Jane Bennett, *The Enchantment of Modern Life: Attachments, Crossings and Ethics* (Princeton: Princeton University Press, 2001).

[87] Belief is what can be practically examined, and faith is the sum of practical experience of the usefulness of this world, the underworld, and the heavens. The meaning of faith in pre-Christian views allows a post and pre-dialectic worlding where examination and experimentation are not opposed to belief. Janīna Kursīte, "Latvieša pasaules uztvere un ticība." On virtue: Kursīte, "Dainu kodekss," 12–13, 534.

[88] Kimmerer, *Braiding Sweetgrass,* citing Gary Nabhan.

[89] Bennett, *Vibrant Matter: A Political Ecology of Things* (Durham: Duke University Press, 2010).

to creation beyond nature, ultimately overcoming the experienced dialectics of nature vs culture and life vs death.

With this story, I, thus, propose to finish here this inquiry of life and nature in Latvian pagan worldings, recognizing within it both the unspoken presence of the stone beginnings, implied by the chthonic undercurrents of the theme, as well as the extensive—time-space-organizing presence of trees that allow the separation of the world of the living and the dead, yet, at the same time, building links between both realms. As such, the stone and the forest in Latvian mythology have shown the possibilities to frame and, most importantly, experience a Dionysian understanding of life and nature beyond and before the dichotomy of life *vs* death and nature *vs* culture, while still retaining an Apollonian structure for the understanding of the world that includes the possibilities of A being not-A, yet upholding a much more complicated frame of reference of the relatedness within.

Here I have tried to demonstrate some of the ways in which the cultural presence of alternate worldings and their durability (even if encapsulated in a fixed form of folkloristic legacy of the past) can be employed for the task of shifting lived ontologies in both experiential and theoretical contexts. While a return in the past is impossible and most probably unnecessary, these ontologies already co-constitute us and remind us of our ongoing relatedness with the world. Moreover, it seems possible, that rebuilding and revitalizing these connections can bring to the fore affections and experienced embeddedness that are often subdued by the frantic rhythms of the technocapitalist lifeworlds we inhabit or silenced by a sorrowful sigh about all that we have lost. What the ethical task of situated knowledges can be, thus, is to provide embedding anchors of meaning and sensefulness that guide through staying faithful to Earth in an immanent, non-abstracted sense. What the practical task of such anchoring would be, is to provide tools to stay true to a post-colonized Earth without despair.

Acknowledgments This article is supported by the ERAF (European Regional Development Fund) Post-doctoral Research Support Program [project Nr 1.1.1.2/VIAA/1/16/001 research application Nr. 1.1.1.2./VIAA/4/20/613, project "Ontogenealogies: The Body and Environmental Ethics in Latvia"] as well as the Fundamental and Applied Research program of the Latvian Council of Science, project "Competing discourses of nature in Latvia and ecological solidarity as a consensus building strategy" (lzp-2020/1-0304).

Bibliography

Alaimo, Stacy. 2010. *Bodily Natures: Science, Environment, and the Material Self.* Bloomington: Indiana University Press.

Barad, Karen. 2007. *Meeting the Universe Halfway. Quantum Physics and the Entanglement of Matter and Meaning*. Durham: Duke University Press.

Barovskis, Ingus. 2015. *Htoniskā pasaule latviešu folkorā: laiks un telpa* [Chthonic World in Latvian Tales: Time and Space], Doctoral dissertation. Rīga: University of Latvia.

Bennett, Jane. 2001. *The Enchantment of Modern Life: Attachments, Crossings and Ethics.* Princeton: Princeton University Press.

———. 2010. *Vibrant Matter: A Political Ecology of Things*. Durham: Duke University Press.
Brown, Nancy Marie. 2022. *Looking for the Hidden Folk: How Iceland's Elves Can Save the Earth*. New York: Pegasus Books.
Cronon, William. 1996. The Trouble with Wilderness; or, Getting Back to the Wrong Nature. *Environmental History* 1 (1): 7–28.
Dainuskapis.lv, Krišjāņa Barona Dainu skapis [*The Cabinet of Folksongs by Krišjānis Barons*], https://www.dainuskapis.lv/. Accessed 27 Feb 2023.
Deleuze, Gilles. 1983. *Nietzsche and Philosophy*. London: Continuum.
Domanska, Ewa. 2018. Is This Stone Alive? Prefiguring the Future Role of Archaeology. *Norwegian Archaeological Review* 51 (1-2): 22–35. https://doi.org/10.1080/00293652.2018.1553060.
En.lfk.lv. Latviešu folkloras krātuve [*Latvian Folklore Collection]*, "The Cabinet of Folksongs or Dainu Skapis", http://en.lfk.lv/collection/folksong-cabinet. Accessed 27 Feb 2023.
Gaard, Greta. 2014. Indigenous Women, Feminism, and the Environmental Humanities. *Resilience* 1 (3): 86–99.
Gimbutas, Marija. 1983. *The Balts*. London.
———. 1989. *The Lanugage of the Goddess*. San Franscisco: Harper & Row.
Guha, Ramachandra. 2013. Radical American Environmentalism and Wilderness Preservation: A Third World Critique. In *The Future of Nature. Documents of Global Change*, ed. L. Robin, S.S. Sörlin, and P. Warde, 409–426. New Haven: Yale University Press.
Iļģi, "Vissbija", *Album "Saules meita" [Daughter of the Sun],* YouTube. https://youtu.be/O32ZjGfcy4A. Accessed 27 Feb 2023.
Irigaray, Luce. 1985. *This Sex Which Is Not One*. Trans. Catherine Porter. Ithaca: Cornwell University Press.
Kennedy, Greg. 2007. *The Ontology of Trash*. Albany: State University of New York Press.
Krimmerer, Robin. 2013. *Braiding Sweetgrass*. Minneapolis: Milkweed Editions.
Kursīte, Janīna. 1996. *Latviešu folklora mītu spogulī [Latvian Folklore Reflected in Myths]*. Rīga: Zinātne.
———. 1999. *Mītiskais folklorā, literatūrā, mākslā [The Mythical in Folklore, Literature, Art]*. Rīga: Zinātne.
———. 2003. Par Austrumprūsijas un Kurzemes vietvārdu stāstiem [*On Eastern Prussian an Courland's Toponyms*] *Latvijas Vēstnesis* 183. https://www.vestnesis.lv/ta/id/82546. Accessed 28 Feb 2023.
———. 2018. *Dainu kodekss [Codex of Daina]*. Rīga: Rundas.
———. 2020. *Latviešu dievības un gari [Latvian Deities and Spirits]*. Rīga: Rundas.
Kursīte, Janīna. "Latvieša pasaule" [*Latvian World*] (lecture course, OpenMinded, Riga, Latvia). https://open.lv/products/janina-kursite-latviesa-pasaule/.
Kursīte, Janīna. Lecture 1 "Latvieša pasaules uztvere un ticība" [*Latvian worldview and beliefs], Latvian World,* Lecture course, OpenMinded, Riga, Latvia.
Kursīte, Janīna. Lecture 2 "Latvieša laika un vietas uztvere. Svētbirzis jeb "dots devējam atdodas"" [Latvian understanding of time and space. A sacred grove or "you get what you give"], Latvian World, Lecture course, OpenMinded, Riga, Latvia.
LaFauci, Lauren and Cecilia Åsberg. 2020. Is All Environmental Humanities Feminist Environmental Humanities? *Seeing the Woods: A Blog of the Rachel Carson Center for Environment and Society*, July 6, 2020. https://seeingthewoods.org/2020/07/06/is-all-environmental-humanities-feminist-environmentalhumanities/.
Laime, Sandis. 2009. The Sacred Groves of the Curonian Ķoniņi: Past and Present. *Folklore (Estonia)* 42: 67–80.
Laime, Sandis, and Beatrise Reidzāne Reidzāne. 2018. Latvian Folklore Studies and Mythology. In *Latvia and Latvians: Collection of Scholarly Articles*, vol. 2, 89–125. Riga: Latvian Academy of Sciences.
LSM.lv. "Mežu platība Latvijā turpina pieaugt – pērn tie bija 52%" [The Area of Forest Keeps on Increasing Last Year it Was 52%]. https://www.lsm.lv/raksts/zinas/latvija/meza-platiba-latvija-turpina-pieaugt%2D%2Dpern-tie-bija-52.a325062/.

Marder, Michael. 2021. *Dump Philosophy: A Phenomenology of Devastation*. London: Bloomsbury.
Marler, Joan. 2022. Baltic Archaeology, Cultural History, Ancient Lithuanian Symbolism, Old Europe, and the Archaeomythology of Marija Gimbutas. *Archaeologia Lituana* 23: 10–33.
Merchant, Carolyn. 1990. *The Death of Nature. Women, Ecology and the Scientific Revolution*. San Francisco: Harper & Row.
Muktupāvela, Rūta. 2012. The Mythology of Ethnic Identity and the Establishing of Modern Holy Places in Post-Soviet Latvia. *The Pomegranate* 14 (1): 69–90.
Neimanis, Astrida. 2017. *Bodies of Water*. London: Bloomsbury.
———. 2018. Posthuman Phenomenologies for Planetary Bodies of Water. In *A Feminist Companion to the Posthumanities*, ed. Cecilia Åsberg and Rosi Braidotti, 55–66. Cham: Springer International Publishing AG.
Nietzsche, Friedrich Wilhelm. 1999. In *Sämtliche Werke: Kritische Studienausgabe*, ed. Giorgio Colli and Mazzino Montinari, vol. 4. München: Deutscher Taschenbuch Verlag de Gruyter.
Petr, Jehlička, Grīviņš Miķelis, Visser Oane, and Balázs Bálint. 2020. Thinking food like an East European: A critical reflection on the framing of food systems. *Journal of Rural Studies* 76: 286–295. https://doi.org/10.1016/j.jrurstud.2020.04.015.
Povinelli, Elizabeth. 2016. *Geontologies: A Requiem to Late Liberalism*. Durham: Duke University Press.
Radomska, Marietta. 2016. *Uncontainable Life: A Biophilosophy of Bioart*. Linköping: Linköping University Electronic Press.
Rasmussen, Rune Hjarnø, "*Nordic Animism*." YouTube channel. https://www.youtube.com/@NordicAnimism/videos.
Sauka, Anne. 2020a. The Nature of Our Becoming: Genealogical Perspectives. *Le foucaldien* 6 (1): 1–30. https://doi.org/10.16995/lefou.71. [Note: In 2022, Le foucaldien relaunched as Genealogy+Critique.].
———. 2020b. *Life in Process: The Lived-Body Ethics for Future*", Religious-Philosophical Articles [Reliģiski – filozofiski raksti] 28: 154–183.
———. 2020c. A Lack of Meaning?: Reactive Nihilism and Processual Materiality. *Approaching Religion* 10 (2): 125–140. https://doi.org/10.30664/ar.91788.
———. 2022a. A Bite of the Forbidden Fruit: The Abject of Food and Affirmative Environmental Ethics. *Open Philosophy* 5 (1): 281–295. https://doi.org/10.1515/opphil-2022-0200.
———. 2022b. Selfhood in Question: The Ontogenealogies of Bear Encounters. *Open Philosophy* 5 (1): 532–550. https://doi.org/10.1515/opphil-2022-0211.
———. 2022c. Becoming Self: A Legion of Life in a Culture of Alienation. In *Normalitāte un ārkārtējība filosofiskā skatījumā [Normality and Exceptionality in Philosophical Perspective]*, ed. Kitija Mirončuka, 25–46. Rīga: LU Akadēmiskais apgāds.
———. 2022d. Beyond the Skin Line: Tuning into the Body-Environment. A Venture into the Before of Conceptualizations. *The Polish Journal of Aesthetics. Listening and Polyphony. Philosophy, Aesthetics, Art* 64 (1): 162–181.
———. 2023. Breaching the Dialectic with Situated Knowledges: The Case of Postsocialist Naturecultures. *Polish Journal of Aesthetics. Postsocialist Affectivity and Intersectional Feminism* 68 (1): 35–56.
Schoeller, Donata, and Sigridur Thorgeirsdottir. 2019. Embodied Critical Thinking: The Experiential Turn and its Transformative Aspects. *philoSOPHIA: A Journal of Continental Feminism* 9 (1): 92–109.
Schwartz, Katherina. 2007. The Occupation of Beauty: Imagining Nature and Nation in Latvia. *East European Politics & Societies* 21 (2): 259–293. https://doi.org/10.1177/0888325407299781.
Scott, James. 2017. *Against the Grain: A Deep History of the Earliest States*. New Haven: Yale University Press.
Škof, Lenart. 2024. "Elements of Slovenian Indigenous Religion", in this volume.
Stauga, Jolanta. 2011. *Dzīvnieki latviešu folklorā: Mitoloģiski maģiskais aspekts (pēc latviešu tautasdziesmu un mūsdienu folkloras materiāliem)* [Animals in Latvian Folklore: Mythological Magical Aspect (After the Materials of Latvian Folksongs and Modern Folklore)], Doctoral dissertation, Riga: LU.

Strand, Sophie. 2022. *The Flowering Wand: Rewilding the Sacred Masculine*. Rochester: Inner Traditions.

Taivāns, Leons Gabriels. 1992. Baltu reliģija un kristietība (tipoloģiskas rekonstrukcijas principi). *The Baltic Religion and Christianity (Principles of Typological Reconstruction). Ceļš* 44: 18–32.

Yunkaporta, Tyson. 2020. *Sand Talk: How Indigenous Thinking Can Save the World*. New York: HarperCollins.

Zaroff, Roman. 2019. Some Aspects of Pre-Christian Baltic Religion. In *New Researches on the Religion and Mythology of the Pagan Slavs*, ed. Patrice Lajoye, 183–219. Lisieux: Lingva.

Anne Sauka (Dr. phil.) is a docent and senior researcher in social philosophy at the University of Latvia. Anne studies materially embedded genealogies of the body and the environment and is especially interested in seeking affirmative ethicalities for environmental humanities. She strives to synthesize first-person experiential approaches (such as embodied critical thinking) with an ontogenealogical approach of (re) visiting mythologies and ontologies that co-constitute lived materialities today. For this, Nietzsche, with his call to seek out the sensefulness of the Earth, has been Anne's central philosophical inspiration, and her research interests include critical genealogy, eco-phenomenology, new materialism, biophilosophy, feminist posthumanism, process ontologies and environmental philosophy. Anne's latest publications can be found here: https://www.researchgate.net/profile/Anne-Sauka/publications Contact: anne.sauka@lu.lv

Chapter 5
The Glacier and Me: Understanding the Human Being in the Anthropocene

Guðbjörg R. Jóhannesdóttir

Abstract In this chapter, I will contemplate a new understanding of the human being that is needed in times of the Anthropocene. An understanding of the human being that is at the same time an understanding of nature, and helps us gain a deeper understanding of the complex and inherently relational reality we live in. When I stand in front of a glacier and sense that both of us, the glacier and me, are threats at the same time as we are under threat, I understand and sense my relations with the forces of nature in a very different way than for example the farmers in Öræfi in Iceland did when they decided early last century to divert a glacial river away from their lands using only picks, shovels, horses and their own hands (M. Jackson, *The Secret Lives of Glaciers* (West-Brattleboro, Vermont: Greenwriters Press, 2019), chapter 4). We are not opposite each other anymore, me and the glacier, rather we are melting together. This melting is both happening in a literal sense as a result of the climate crisis, as well as metaphorically as we are understanding better and better how the borders between humans and nature are disappearing or melting away, and how they have perhaps never existed. The chapter will address the question of what characterizes this new understanding of the human being from the perspective of relational ontologies and embodied aesthetic knowledge.

Keywords Anthropocene · Relational ontology · Aesthetic experience · Embodied knowledge

What I have been focusing on in my research for the last decade or so is the aesthetic experience of landscape in Iceland, and the value and meaning of such experiences. In my PhD project I focused especially on the landscape types that characterise the

G. R. Jóhannesdóttir (✉)
Department of Arts Education, Iceland University of the Arts, Reykjavik, Iceland
e-mail: gudbjorgr@lhi.is

L. Škof et al. (eds.), *Elemental-Embodied Thinking for a New Era*, Sophia Studies in Cross-cultural Philosophy of Traditions and Cultures 42,
https://doi.org/10.1007/978-3-031-42119-8_5

Icelandic highlands, glacial and geothermal landscapes.[1] One of the things that I found through that study was that aesthetic experiences, for example the experience of touching and being touched by a glacier, can provide people with the opportunity of gaining deep rooted ethical knowledge of their place in relation to nature. It is this knowledge that I would like to explore here in the context of the situation we are in as humankind in the age of the Anthropocene: a concept that refers to the fact that human beings have become a geological force. We are in an overwhelming situation, and I wonder whether the glacier can help us? Can the glacier help us gain the new understanding of what it means to be human that we need to be able to find our footing in this situation?

As I said, we are in an overwhelming situation where we are confronted with extremely urgent tasks, but it seems like we are frozen in our steps. Why are we so frozen?—and what does it take to be able to move? Being informed about the devastating effects of global warming is not enough to get us moving. I agree with Timothy Morton who claimed in a lecture he gave in Iceland that the shock-mode we have been in lately in reaction to the endless flood of information, the "yelling about climate change," is a part of our failure to respond to the climate catastrophe that is occuring—the shock mode is partly the reason why we are frozen in our steps.[2] Like the Norwegian psychologist Per Espen Stoknes has pointed out, most people respond to the shock therapy by going into denial, giving up and putting their fingers in their ears.[3] So our problem is not that we don't have enough scientific data to support the facts that we see and experience all around us by now, and it is not that we are not mediating this data widely enough. The problem lies in something much deeper—I think our problem is that we need to be able to integrate a different understanding of what it means to be human—a new self-understanding for humanity. And just like I wouldn't be able to gain self-understanding by reading graphs or books that someone would write about me seen from the outside, we as humanity can't gain understanding of ourselves and our place in the world in that way. We need to experience a new understanding of the human being on our own skin—through our bodies, from the inside out—and that is what I will be focusing on here.

Can the concept of the Anthropocene provide us with some clues about what kind of understanding of the human being we have today and in what directions it could possibly develop? Can the idea of the Anthropocene help us? What does this idea tell us? *Anthropo*-cene—the era of humans—if I'm honest this doesn't sound very good to me. It sounds like this is the era where humans are in charge, they control, they are at the center—and that does not sound too good to ears that are acutely aware of how the idea of the human that controls nature has been leading us

[1] Guðbjörg R. Jóhannesdóttir, *Icelandic Landscapes: Beauty and the Aesthetic in Environmental Decision-making* [PhD Thesis in philosophy] (Reykjavík: University of Iceland, 2015).

[2] Timothy Morton, "You're Not Making Ecological Art Yet" (open lecture organised by the department of fine art at the Iceland University of the Arts and the Institute for philosophy at the University of Iceland), Safnahúsið, Reykjavík, February 2nd 2018). https://vimeo.com/254647013

[3] Per Espen Stoknes, *What We Think About When We Try Not to Think About Global Warming: Toward a New Psychology of Climate Action* (Hartford, Vermont: Chelsea Green Publishing, 2015).

into the impossible situation we are in. However, if we understand the concept of the Anthropocene in this way, then we are misunderstanding the ideas behind it very much. That someone is in control, and dominates, implies duality—here am I, and then there is something outside of me that I control and dominate over. But the concept of the Anthropocene does not describe a duality—it describes merging—as the geological concept that it is, it describes how humans are literally merging with the earth—we are a part of the earth in such a way that in millions of years our imprint will be found in the geological layers of the earth. Then the new geological formation Plastiglomerate[4] will testify to our existence among other things.

So there are two sides to the concept of the anthropocene—the side that shows humans in control of nature on the one hand and on the other hand the side that shows humans as part of the earth in a very literal way. But this is nothing new. These two sides have characterised our understanding of the human being for ages, as Páll Skúlason wrote in his paper "Man in the state of nature": "from ancient times the debate on the position of man in the state of nature has depended on this strange loop that nature and the human spirit take around each other [...] should we see ourselves as a tiny fragment of the material world of nature, absolutely dependent on how its forces play with us? Or should we consider that with the power of our own spirit we can free ourselves from nature, rise above all material reality and be free to think the world according to our own wants?"[5] Isn't it time that we step out of this strange loop that assumes two opposite alternatives and no middle-way inbetween? The fact that the situation we are in calls for urgent action, somehow calls us to seriously shake things up—*even to turn everything around at once*. And in which way should we turn? Should we expand and deepen the understanding of the human being that is in control of nature or should we expand and deepen the understanding that we are part of and intertwined with nature? In my mind the first option is very dangerous—and in fact it is the root of the crisis we are already in. The other option is to deepen our understanding of how we merge with the earth. Even though this understanding emerged a long time ago—ecologically we have known for a long time what our place within the ecosystem, is—we still haven't taken this understanding to heart—we still haven't understood how we merge with the earth in a deeper way than ecologically and geologically, and we have still not embraced this understanding all the way through and into our bodies.[6]

[4] Geologists have now identified and named this new type of stone formed of melted plastic, beach sediment, basaltic lava fragments and organic debris, see: Patricia L. Corcoran, Charles J. Moore og Kelly Jazvac, "An Anthropogenic Marker Horizon in the Future Rock Record,"*GSA Today* 24, no. 6 (2014): 4–8.

[5] Páll Skúlason, "Maðurinn í ríki náttúrunnar [Man in the state of nature]," in *Náttúrupælingar[Reflections on Nature]* (Reykjavík: University of Iceland Press, 2014), 81–88. My translation.

[6] When I say "we" I'm referring to the western culture that has been disconnected from nature in a way that many other cultures around the world have not.

In his above-mentioned lecture, Morton pointed out how badly we are doing in facing up to the climate crisis: "We are having trouble living the fact of it."[7] It is as if we are stuck in responding either with the thought that we are not living in times of climate change yet or with the thought that we have a choice—that there is a possibility that climate change is not happening. We are stuck there frozen in our steps because we can't live the fact of it. And what is that? Morton rightly pointed out that: "Part of living something for real involves being able to feel it out on the inside—being able to imagine it."[8] And what does it mean to feel something out on the inside? If one truly has ingrained knowledge of something then one can feel it from the inside. This happens when we try to visualize a place we know well, or when we lose the thread of what we wanted to say and start looking for this felt sense[9] of the thread that we know is there. If we don't have the knowledge then we don't have this inner thread to look for. This seems to be the case with global warming, we have not been able to take this knowledge to our hearts—to the core of our being.

So how can we start living this fact?—how can we feel global warming from the inside and find appropriate responses? And how can we feel from the inside a new understanding of what it means to be human that we can build our responses on? To answer these questions I am going to examine my own relationship with the glacier and explore whether this examination can tell me something about feeling knowledge out from the inside. This phenomenological approach is partly inspired by Claire Petitmengin's writings on anchoring in lived experience as an act of resistance to our current global crisis[10] and by M. Jackson's writings on glacier-human relations.[11] As Jackson points out glaciers are "iconized with climate change, but as this transpires, everything else about a glacier seems to fall away," including the meaning we can find in human interaction with glaciers. Attending to and exploring our lived experiences of our relations with the more than human is one way of anchoring our whole being in the sense of being one with, not separated from, the ongoingness of life on earth. To begin this exploration I will share the first two experiences of coming close to glaciers I ever had. Standing in front of Hoffellsjökull glacial tongue for the first time (this was before a lagoon started forming in front of it) I wrote:

> And then we came up to the glacier and I felt it was pulling me in. The soundscape was so still and beautiful, water streaming, and sounds of tiny bits of ice loosening and then falling into the water. When we came to the edge and I looked up at the iceberg above me and into the crevasse and listened to the still and slow sounds I felt that the glacier was so still and slow and also so infinitely old, ancient and powerful. I was thrown into this feeling all of a sudden; felt my stomach tighten and almost like a lump in my throat.

---

[7] Morton, "You're Not Making Ecological Art Yet," min. 14.39

[8] Ibid, min. 22.06.

[9] This is a term coined by Eugene T. Gendlin, see Gendlin, Eugene T. Focusing. New York: Bantam, 1981; Gendlin, Eugene T. Experiencing and the creation of meaning: A philosophical and psychological approach to the subjective. Northwestern University Press, 1997.

[10] Claire Petitmengin, "Anchoring in lived experience as an act of resistance." *Constructivist Foundations* 16, no. 2 (2021): 172–181.

[11] M Jackson, *The Secret Lives of Glaciers*.

> A year later I experienced for the first time to be on the ice sheet on the top of the glacier, when I travelled with a superjeep across Vatnajökull. There I experienced other layers of being with the glacier. Crossing the middle of the glacier I wrote:All of a sudden the glacier appeared with all its deep crevasses and I was a little startled, my heart skipped a beat – indescribable beauty and power.
> Came up on the glacier and expanse, expanse, broad horizon....endlessness.
> What is underneath us now? What an enourmous mass of ancient ice.
> White, white, endless white, becomes whiter and cleaner as we go higher.
> The horizon white lines.
> Strange feeling, weird atmosphere. No words.
> My heart skipped a beat when we crossed a little crevasse.
> The clouds, the glacier, the sky, white, white, white and blue.
> I want to hear the silence outside.
> I'm so high up and there is so much mass underneath me. Endlessness in all directions, up, down and to the sides.
> White merges with white, glacier with clouds.
> I would get lost in the fog. Mysterious energy here. The white white light. I wanted to dissappear into the glacier. Total silence, total silence, totally nothing, total calmness, total energy—the white light that immerses you gives you a strange energy—white in through the eyes, white into me.
> 1770 meters high and up in the clouds. We are up in the sky.

The important thing here is how my bodily felt sensing appears in these examples. In both cases I was somehow overwhelmed by bodily sensing—when I looked up at the glacial wall above me my stomach tightened and I got a lump in my throat, almost paralyzed by this ancient energy, felt this instant closeness, being eye to eye with this ancient, still, slow, powerful being. It is important here that I didn't feel I was up against an object, a thing, but rather I felt I was *face to face* with a being—something or someone I could communicate with in some manner.[12] The same occurred on the top of Vatnajökull, I'm sensing with my whole body this massiveness below my feet. Then I write: "I want to hear the silence outside." At this moment this bodily sense had become so strong that I felt the need to come closer to this phenomenon that I was sensing through the windows of the car. I wanted to stop the car and get out, I felt that the car and the sounds that came with it somehow didn't belong in this world. I felt conflicted, feeling that we didn't belong there with this machine. There was some feeling of agitation connected to the car that refers to a certain ambiguity in these experiences that I think is important to explore further.

This agitated feeling has to do with our reliance on machines and technological knowledge. The driver who drove us over Vatnajökull didn't see the crevasses on the glacier, we often didn't see anything except white fog and snow blizzard out the window, but the driver knew where the crevasses were because he had them marked

[12] In M Jackson's book *The Secret Lives of Glaciers*, a whole chapter (chapter six) is dedicated to exploring how people she interviewed in Iceland describe how they sense the glacier as a living being, so this experience of mine is not an isolated one. Many indigenous cultures see the whole natural world and phenomena within it as living beings, for example the Maori in New Zealand. This world view of the Maori has now been integrated into New Zealand laws where the river Whanganui has been granted legal personhood. See Abigail Hutchison, "The Whanganui river as a legal person," *Alternative Law Journal* 39, no. 3 (2014): 179–182.

on a GPS system that told him where the car was on the glacier. This can be seen as a representation of the side of the Anthropocene that has to do with the understanding of the human being that we have been stuck within. The old, deep rooted western understanding of the human being as a rational being that is totally different from all other beings; the human being that is on top of creation with its scientific knowledge and technology that makes us able to control nature. But like I said before, the Anthropocene also has another side to it, a different understanding of what it is and means to be human, an understanding of the human being as totally intertwined with the earth. This is the ambiguity that I felt so clearly and in such a strong bodily felt way there on top of the glacier. And what I sense when I go back to this is how filled with emergency the moment we are in right now is. We live in a time where we can go in both directions, and I think we need to go deeper into and understand better what it is to be intertwined with the earth, not just geologically but also spiritually and intellectually, and then perhaps we can see how these two sides of the Anthropocene can also intertwine and support each other.

What does it mean to melt into the earth? In a certain way we have already come to understand that long ago, in terms of the ecological understanding that developed in the last century.[13] But this understanding does not seem to be enough, this melting that we need to understand now, does not just involve hugging trees, it is about something much deeper than that. It does not only involve understanding that we, me and the glacier, are a part of the same process, the same ecosystem, rather it now also involves understanding that we, me and the glacier, are both processes, we are not objective units, we are processes and combinations of materials that come together in different ways in relation to the situation that the processes are in. And we are the same kind of processes in many ways, perhaps even the same process. Me and the glacier are for example both made of water, both of us are at the same time solid and liquid forms, we are processes, forces and powers like anything else on the planet. We are processes that influence and are influenced by our environments. We are also processes that are a threat and are threatened at the same time. We have totally melted into the glacier in the way we have become both vulnerable and threatening at the same time. Being a threat and being threatened.

This sense of being vulnerable and threatening is stirred in us every time we see a graph that shows climate catastrophes and news about storms, floods and forest fires, but we can't bare to be with it, to be able to become aware of it and let it move, we haven't been able to live this feeling from the inside like Morton talked about. The reason why we don't manage or don't want to live this feeling from the inside is that it is so paralyzing. Perhaps this is somehow similar to when we as individuals resist taking in the fact that we are mortal; in this case it is our resistance to acknowledge the fact that humanity as a whole is mortal. This is a paralyzing sense of not

[13] In a way, this ecological understanding was not something new, as it can be seen as a remembering of the ancient world views that still prevail in some indigenous cultures, where a separation between humans and the larger living system of nature is unthinkable.

being able to do anything, we the public are so powerless in relation to authorities and big corporations of the world that seem to be resisting action, resisting real and serious change. What can we do about this paralyzing feeling? What can we do to move out of these frozen steps we are in? Can the glacier as a process like us teach us something? If we understand that the glacier is a process like we are, can we then look at what kind of process the glacier starts in this situation and mirror ourselves in that? What does the glacier do in this situation we are together in—being a threat and being threatened at the same time? It just melts. It is not spending its time wallowing in the feeling of experiencing itself as a threat that is threatened. It doesn't even experience itself as a threat, doesn't experience anything, just melts, gets paralyzed and melts. We could say that this is exactly what we are doing; not experiencing ourselves as a threat, not being able to take in and live the fact that we are a threat that is threatened, and meanwhile we are just paralyzing and melting. But the glacier does something more than melt, it takes this melting into its inner landscape and changes it, transforms it through the melting. We also need to take this melting we are experiencing together, us and the glacier, into our inner landscapes. By doing that we can go further in doing like the glacier: It clears off the old landscape and creates a new one. As humanity we are in that phase, where the old comes crumbling down suddenly, like happened in the #metoo-revolution and other big social revolutions. Something happens that turns everything around, sometimes through big and noticeable revolutions, sometimes more silently. Little by little more things come to the surface that have been frozen still, quiet, silenced, and overwhelming, like what came up in the #metoo-revolution. Suddenly everything is up on the surface where it can melt, and then evaporate and leave a new landscape like the glacier does. If we manage to clear away our old inner landscapes, our old understanding of the human being, with more revolutions like this, then we can be optimistic.

Perhaps it is time for nature to revolt like women did in the #metoo-revolution? As eco feminist theories have pointed out for a long time, there are many things in common with the situation of women and that of nature in our patriarchic systems.[14] Perhaps we can see nature's increasing loudness with her more and more frequent storms, forest fires, frosts, floods, earthquakes, volcanic eruptions and viruses as her way of saying enough is enough? The question is just whether we are ready to listen, or how loud she must scream to really shake us up and turn everything around.

Whether this revolution in our understanding of humans and nature is happening slowly and quietly or if something happens that turns everything around in one moment, I think there are enough signs showing that it will happen. What we are gradually sensing more and more is that what happens in such an intense and focused bodily perception that leads to an aesthetic experience of melting into the landscape, is in fact always happening, every moment, just in a subtler way. When I sit and stroke my cat, or when Donna Haraway's dog licks her face there are no

[14] See Karen J. Warren, *Ecofeminism: Women, Culture, Nature* (Bloomington: Indiana University Press, 1997).

boundaries between us and animals.[15] When I sit by the computer there are no boundaries between me and the keyboard, it has become an extension of me. We are not the separated individual entities that western culture has told us we are. Morton gave a good advice to get used to this thought: he said he had gotten into the habit of thinking about objects as if they were water. This is the understanding of the human being that needs to settle into every cell of our bodies. Being on the glacier is probably one of the most effective ways of gaining this understanding, but the next best way, if we can't go to the glacier, is to gain this deeper understanding through artists' and poets' mediation of this earthly connection to the world. We already know through our rationality what and how the glacier is as a measurable phenomenon, what we must focus on now is understanding our relation to the glacier through our senses, through our imagination. Imagination is the root of rationality that makes us capable of envisioning responses, putting ourselves imaginatively into a situation—that is something we cannot calculate. We need to be able to see ourselves in the situation and see possible responses. The word responsible—means being able to respond. When this understanding settles into every cell of our bodies and becomes accessible to our imagination we can really take in, and melt into, and live that fact that we are a threat that is threatened like the glacier— and then we can start responding appropriately; we can continue to be a threat that can clear out the political and cultural landscapes of the patriarchal past and create new landscapes of hope for the future.

Bibliography

Corcoran, Patricia L., Charles J. Moore, and Kelly Jazvac. 2014. An Anthropogenic Marker Horizon in the Future Rock Record. *GSA Today* 24 (6): 4–8.

Haraway, Donna. 1991. *Simians, Cyborgs, and Women: The Reinvention of Nature*. London/New York: Routledge.

———. 2008. *When Species Meet*. Minneapolis: University of Minnesota Press.

Hutchison, Abigail. 2014. The Whanganui river as a legal person. *Alternative Law Journal* 39 (3): 179–182.

Jackson, M. 2019. *The Secret Lives of Glaciers*. West-Brattleboro: Greenwriters Press.

Jóhannesdóttir, Guðbjörg R. 2015. *Icelandic Landscapes: Beauty and the Aesthetic in Environmental Decision-making* [PhD Thesis in philosophy]. Reykjavík: University of Iceland. https://skemman.is/handle/1946/20813.

Morton, Timothy. 2018. *You're Not Making Ecological Art Yet*. Open lecture organised by the department of fine art at the Iceland University of the Arts and the Institute for philosophy at the University of Iceland, Safnahúsið, Reykjavík, February 2nd 2018. https://vimeo.com/254647013.

Petitmengin, Claire. 2021. Anchoring in Lived Experience as an act of Resistance. *Constructivist Foundations* 16 (2): 172–181.

[15] See Donna Haraway, *Simians, Cyborgs, and Women: The Reinvention of Nature* (London & New York: Routledge, 1991); *When Species Meet* (Minneapolis: University of Minnesota Press, 2008).

Skúlason, Páll. 2014. Maðurinn í ríki náttúrunnar [Man in the state of nature]. In *Náttúrupælingar[Reflections on Nature]*, 81–88. Reykjavík: University of Iceland Press.

Stoknes, Per Espen. 2015. *What We Think About When We Try Not to Think About Global Warming: Toward a New Psychology of Climate Action*. Hartford: Chelsea Green Publishing.

Warren, Karen J. 1997. *Ecofeminism: Women, Culture, Nature*. Bloomington: Indiana University Press.

Guðbjörg R. Jóhannesdóttir is an associate professor at the Iceland University of the Arts. Her current research within phenomenology and aesthetics focuses on human-environment/body-landscape relations and processes, and their role in human thinking and understanding.

Chapter 6
How Deep Is Your Love or, How Thinking as Such Defies the Weight(s) of Imperialism

Treena Balds and Timothy Morton

Abstract Elemental and embodied thinking disrupts the master–slave template by formulating thinking in such a way that it does not require an object. In fact, it is the only way to think at all. Elemental thinking allows humans to respect the voices and subjectivities of all the planet's inhabitants and to apply this relational posture to interactions between the various races that constitute our species. The rise of fascisms planet-wide is, like the growl of an injured predator, a signal of the imminent collapse of the master–slave template as the dominant format, one which also stands-in for other toxic binaries such as active–passive and human–animal. Imagining thought without this template is ecologically vital. We find a paradigm for this new thinking in what some might consider an unlikely place: conventional (that is to say, non-Gnostic) forms of Christianity.

Keywords Charles Darwin · Colonialism · Elements · Evolution · Martin Heidegger · Imperialism · Mathematics · Racism · STEM · Object-oriented ontology

Quite seriously, the topic of this collection is not only a valuable way to think—elemental and embodied thinking *is the only way to think at all.* In scandalous opposition to traditional conceptions of thought, we apply to the physical the concept of sentience. This act requires an interaction between a lifeform and its environment that we take to be the basis of the term "elemental": a word whose valence is neither subjective nor objective. More precisely, the elemental is a term that establishes the ground out of which a reification such as "subject" and "object" can arise, but not necessarily.

T. Balds
Department of English, Rice University, Houston, TX, USA

T. Morton (✉)
Starter Academy, Montego Bay, Jamaica
e-mail: tbm2@rice.edu

L. Škof et al. (eds.), *Elemental-Embodied Thinking for a New Era*, Sophia Studies in Cross-cultural Philosophy of Traditions and Cultures 42,
https://doi.org/10.1007/978-3-031-42119-8_6

By "elemental" is meant what Tibetan or Galenic medicine and Yoruba spirituality meant by "earth," "wind," "fire," and so on. What is elemental is environmental not because it "surrounds" or even "permeates" an object or a subject (or both) but because it is the basic modality of this environmentality. Martin Heidegger describes this as Dasein and then localizes it in the human being (and eventually, alas, in the German human being first and foremost). But the argument in *Being and Time* simply won't stand up to this localization.[1] Dasein is non-anthropocentric to its core, and moreover Dasein implies the existence of what Graham Harman calls "objects"; to wit, entities that are not defined by their appropriation or indeed by their appropriability at the hands of other entities.[2]

This essay addresses the politics of the idea of "elemental-embodied thinking." We begin with the idea that "elementality" has nothing to do with subject versus object.[3] What is special about thought's being "elemental" is that this suggests that there is a phenomenology of "thought" at every point along its pathways. This phenomenology does not depend on "subject" and "object," so that what is elemental is not "subjective." We may, for instance, assume that Galenic terms such as "water" mean "watery" and that this implies a subject experiencing an object called water. But this is not to have said anything yet. We must now explain terms such as "experience," "object," and "called."

What is better than this merely semantic approach is to realize that phenomenology is the "logic of phenomena," that is to say, the way they arise. This way is elemental through and through, because when the wind stirs the leaves in a tree, the wind does not exhaust the leaves like an extractive corporation exhausting the oil in the Permian Basin in north Texas. The wind "accesses" the tree in a way that is not only similar to the way we access a painting by Leonardo da Vinci in the Louvre: it is isometric with this. The wind lets the leaves speak through it, and by allowing them a voice, accords them a subject position in the relation.

At no stage is the elemental excluded in sequences formed by the movement of the wind, its disturbance of leaves in the tree, the pressure wave in the ear canal, the transduced signals in the brain analyzed as sound, and the arising of a memory associated with that sound. The "quality" (to use a loaded term) of these events *is exactly what they are*, not some accidental decoration.

The other term in the title of this collection also has to do with phenomenology not as "subjective experience of…" but as the foundational "how" of arising. What is "embodied" is one's whole being as a lifeform in a biosphere, but this is also because any thought whatsoever is embodied. As Edmund Husserl observed, an idea (like a tweet or a meme) could never be something floating in a void. Logical propositions cannot simply be symptoms of healthy brains, because healthy brains require science to discern them, which is based on logic, whose propositions are

---

[1] Graham Harman, *Heidegger Explained: From Phenomenon to Thing* (Chicago: Open Court, 2007).

[2] Graham Harman, *Tool-Being: Heidegger and the Metaphysics of Objects* (Peru, IL: Open Court, 2002).

[3] Timothy Morton, "Elementality," in *Elemental Ecocriticism: Thinking with Earth, Air, Water and Fire*, eds. Jeffrey Cohen and Lowell Duckert (University of Minnesota Press, 2015), 271–285.

symptoms of healthy brains.[4] This unsustainable circularity collapses as soon as we acknowledge that to be an idea at all is to be embodied, whether traced along neural pathways inside the lifeform's brain or inscribed "outside" its "body" in words printed on paper or in pixels or in air.

A thought, to put it more simply, *is a feeling*.

There is another way of getting at this entirely, a way that is even more fundamental than what we have argued above. This has to do with the relationship between the future and the past in the phrase "elemental and embodied thinking." For the sake of brevity, let us swap the "feeling" for this phrase. The phenomenology of a feeling is that it is an idea from the future. One goes to therapy because one is having a feeling that one can't explain. Naturally this feeling is a symptom of something that happened "in the past." But it is also happening right now, and one does not yet understand it. Once one has words for it, it might change its quality, perhaps diminishing as one gains some sense of its embodiment, its elementality, its phenomenology. One pays a therapist because one knows that sometime in the future, one's feeling will be articulated as an idea. A feeling is an idea from the future. Therefore, having new ideas requires that one is obliged trust one's feelings, because feelings are ideas from the future.

When we started to write this essay, Treena Balds confessed a peculiar weakness to Timothy Morton: her difficulty telling her left from her right. Morton immediately assured her that far from being a weakness, being unable to tell right from left was a "superpower." Incidentally (or not), it also came out that Morton considered his special weakness to be his name, Timothy, the most popular meaning of which is "timid," "fearful," "lacking courage." The fact that an "incapacity," "disability," "failing," or however one might name it might be unusual and might allow for feelings–thoughts that were different immediately struck Morton as important in the best possible way. If feelings are always futural, then intrinsic to them is the ability to excite affective responses not hidebound by concepts. A feeling always defies the conceptual frame in which it is seen to arise, just as an album by one's favorite band always defies the labels in the record store that the manager and the shoppers associate with it. Things always overflow their names and concepts, because things are made out of time, and thinking–feeling (including how the wind stirs the leaves) can grasp profoundly temporal things only as effectively as do wet hands trying to grab the soap in the bath.

What is new about being unable to tell right from left is that proprioception drives how space and time are measured. We count things on our fingers; we adjust our bodies in one direction based on a sense of hyperextension in the opposite direction. But space and time are not reducible to measuring. One finger pointing in the air is not the same as the number one. The way one carries one's body up a hill is not one with the gradient of a hill. With such radical and perhaps disturbing thoughts (disturbing in a good way, we feel), Immanuel Kant invented phenomenology, the

[4] Edmund Husserl, "Prolegomena to All Logic," in *Logical Investigations*, trans. J.N. Findlay, ed. Dermot Moran (London: Routledge, 2008), 1.1–161.

"science" of elemental embodiment, by insisting that how we access a thing is what that thing "is," for us. Although there is a lot riding on the idea that this was at all "simple": in its ramifications, Kant "simply forgot" to say that reasoning and measuring were not the only ways of accessing things. The wind stirring the leaves also fails to appropriate those leaves just as much or as little as a human being measuring them. It is just that what is neglected in anthropocentric philosophy is this basic fact. What properties of the leaf escape the wind are copious, and this seems obvious to us humans. Yet, many consider Kant's concept of the noumenal aspect of a thing—the thing in itself—to be mysterious or dubious. All it does, though, is simply record (or stand in for) an excess: that in the thing which is beyond a human's capacity to sense or, through any scientific manoeuvring, to calculate.

And why does anthropocentric philosophy neglect all this? One word: colonialism. Six words: colonialism and its master–slave logic.[5] In this essay, we wonder aloud whether it might not be the case that such thinkers are resistant to the concept of the thing-in-itself because of a colonial insistence on a thing's lack of subjectivity—a refusal to grant subjectivity to anything except a human, even to a human of a peculiar kind.

What colonialism does is to measure the world according to a device or set of devices that the colonizer sets up as the one true measurement technology. When this has happened for long enough or enough times, the data gathered by the measurement technology is mistaken for reality. It is no wonder then that a society still reeling from the five-hundred-year legacy of European settler colonialism and slavery would have created the educational regime called STEM, an acronym for "science, technology, engineering, math"—an acronym that in the guise of four pseudo-things is actually saying just one thing. This one thing is not math, but calculation.

The difference between math and calculation is the difference between pointing a finger in the air and the number one. This difference is, as Kant spookily observed, a crack in the universe that one cannot point at directly, a transcendental crack that he rather wonderfully and poetically embodies as "the unknown = X," a phrase that is elemental and embodied indeed.[6] Furthermore, to reduce mathematics to calculation is to contribute to a reality in which some people are already having to calculate all the time, that is to say, way too much, just to survive. If I have to perform complex risk assessments every time I walk down the street so as not to be shot dead by a police officer or by a member of the public—for that matter, a public often empowered by "make my day" laws that allow for murder on the basis of observing me to "put a foot wrong," say by stepping on their front lawn—I am already doing STEM outside the classroom on a daily basis and my life is at best miserable and at worst hell. "Hell" in this sense is precisely a subjection to another's measurement regime.

[5] Denise Ferreira da Silva, *Toward a Global Idea of Race* (Minneapolis: University of Minnesota Press, 2007).

[6] Immanuel Kant, *Critique of Pure Reason*, tr. and ed. Paul Guyer (Cambridge: Cambridge University Press, 1999), 143.

If one is "on" someone else's time, just their time, all the time, that is called oppression.

STEM is empowering only if it is tied firmly to an education that insists on feelings, that is to say, on actual mathematics, embodied for example in English Literature. In short, STEM is empowering only when it is infused with the aesthetic and artistic principles advocated in the STEAM paradigm, in which the inserted A represents the arts. Or, even more radically, the humanities and arts are regarded as the basis for science as such, as they were in the original conception of a medieval European university: the "trivium" of grammar, logic and rhetoric that enabled the "quadrivium" of the sciences.[7]

As many are beginning to observe, the links between math and literature studies are manifold. This is because their phenomenology is identical. It is often said that math is about appreciating beauty, and as Kant argued in the text that grounds the reality of the "unknown = X," that is to say, synthetic judgments a priori (Kantian for "feelings"), beauty is the "feel" or the "ring" of "true" without its content, or as we like to say, beauty is truthfeel.[8] Beauty, in other words, is the phenomenology of "true."

To explore how both literature and mathematics perform beauty (truthfeel) as phenomenology, consider the literary concept of a poem and the mathematical concept of a theorem. Each is an object or phenomenon that, by embodying truth, contains within itself the path (that is, the *phenomenology*) of its own construction during the creative literary or mathematical process.

This path, hylomorphic in its repose in or *as* the object itself, can be excavated and laid bare via the demonstrative processes of explication and proof. A proof, like a literary explication, is a reverse engineering of the progress of creativity toward a perfectly rendered experience on the one hand or a powerfully explanatory theorem on the other. The artist constructs each object by assembling necessary and sufficient conditions for its existence, by pursuing a method that "works out" the most accurate path to its realization. That the path runs inexorably into the object is the very performance of truth, whose effect is akin to what Henry James refers to as the "dreadful done."[9] And the artist or mathematician does not so much perform this truth but witness it by touching it with the receptors sensible to what it performs.

A first encounter with the poem or theorem is a kind of attunement or listening that produces the truthfeel, a synthetic judgment a priori or an intuition of truth. Elemental thought accesses the phenomenology of "true" as that self-same process of assembling conditions that are necessary and sufficient for an object to arise, but in a fashion that lays bare the way they arise. For in courses on proof writing, mathematicians learn that reversing the process of "working out" a problem produces all

---

[7] Sarah Hart, "The Wondrous Connections Between Mathematics and Literature," *New York Times*, April 7, 2023, accessed April 14, 2023. https://www.nytimes.com/2023/04/07/opinion/the-wondrous-connections-between-mathematics-and-literature.html?searchResultPosition=1

[8] Immanuel Kant, *Critique of Judgment* (**).

[9] Henry James, "Preface," in *The Spoils of Poynton*, ed. David Lodge and Patricia Quick (New York: Heinemann, 1908), 126.

the required contents of a sound proof.[10] Similarly, literary scholars prove the truth of literary objects by explications and analyses that exhibit the components and methods of their construction. The final peroration or *quod erat demonstrandum* are insistent gestures toward the beauty revealed as that phenomenological process which culminates in (the) truth (object).

Having established beauty as the phenomenology of "true," what then is the phenomenology of beauty? The Greek word *mathēsis* gets at this phenomenologically quite beautifully, if the reader will pardon the recursion. *Mathēsis* does not *mean* calculating but *grounds* it. Why are we calculating at all? Because of *mathēsis*. *Mathēsis* means "communing with" or "growing accustomed to" or, and this is the preferred term here, "vibing with."[11] The Tibetan term for "growing accustomed to" is göm, which means meditation—in fact, the highest form of meditation known as Dzogchen ("The Great Perfection"), whose basic format is called "nonmeditation," insofar as there is "nothing to see," to wit, nothing to calculate. One has already arrived at the solution, otherwise known as Buddha nature, and the task is simply to get used to that. And this implies allowing one's confusion "to dawn as wisdom," as a liturgy of one of the lineages of Tibetan Buddhism puts it.[12]

What one communes with or grows accustomed to or vibes with is simply a "thing" in its irreducibly mysterious opacity, no matter what phenomena or data it appears to emit. What fascist conspiracy theory offers is a weaponized version of the exact opposite: a preformatted window or magnifying glass or X-Ray machine that renders everything transparent in advance. The ideological trick of a conspiracy theory is to confirm one's prejudices by showing how they reside magically "beneath" appearances, so that when one submits appearance to one's X-Ray (or indeed, one's AR-15 assault rifle) all shall be revealed as anticipated. The pleasure of science, mathematics, and literary criticism resides in resisting this temptation as much as possible.

Which is to say, no matter whether or not truth goes "ting," whether or not the phenomenon called beauty arises, the mathematician and for that matter the literature scholar (or art historian and so on, take your pick of humanities topics), is "vibing on" something, communing with it much like the wind communes with the trees. Whether they are 5 years old and fretting over "one plus one" or whether they are a forty-year-old professional mathematician fretting over Fermat's Last Theorem, they are, while doing *mathēsis*, ready to be "at the wall" and willing for that wall *not yet* to become transparent—a good image for "beauty"—but to remain opaque.

To be okay with vibing on opacity is, in short, exactly the same skill as is taught when a literature professor trains a student to read a poem. The trick of reading a poem is *not to read the poem*. That is to say, what is frightening about QAnon is that

[10] Douglas Smith, Maurice Eggen and Richard St. Andre, *A Transition to Advanced Mathematics*, 5th edition (Boston: Brooks/Cole, 2005), 32.

[11] Martin Heidegger, *What Is a Thing?* (**).

[12] "Supplication to the Takpo Kagyü," tr. Nalanda Translation Committee (Halifax, Vajradhatu, 2011).

this highly disturbing new amalgam of algorithmically collected anti-Semitic memes and tweets such as the Blood Libel is based on "reading" things such as the faces of celebrities as quickly as possible, finding in their winks the eye of the Illuminati for example, and seeing "beneath" appearances to find that some human beings are in fact pedophile lizards who must be exterminated with extreme prejudice.

Humanities classes impede this kind of reading, which is simply following preformatted concepts, the most stupid and readily-available ones still floating around somewhere in the ideological toilet bowl called "the present moment." Humanities resists the end of the world, if by "end of the world" is meant the apocalypse, the violent unveiling of appearances to disclose the reality underneath, the denouement… But a real play ends when the curtain comes down. What is mistaken for reality is in fact a play in which the fascist QAnoner is certain that they are the "goodie" and that they can be sure who is the "baddie." Things fit neatly into preformatted concepts, that is to say, things are neatly titrated into pre-existing measuring cups.

Humanities resists the end of the world because it is a kind of *mathēsis*. And *mathēsis* is most assuredly vibing, to wit, allowing a dataset such as bunch of squiggles on a page to be meaningless. The difference between meaning and meaningfulness is that the latter admits the possibility that "it might just be nonsense." And this is what finally distinguishes what is very valuable about the humanities from STEM. The humanities are the very method required to resist fascism at the most basic and granular level. Furthermore, if "vibing," rather than (just) calculating, were taught more confidently, people who had to calculate way too much just to make it across the street to university—people including Professor Henry Louis Gates who in 2009 was arrested while ascending the (ironically preformatted) steps to his house near to Harvard—people would be able to understand, resist, and generate their own "measuring cups" to frame the world differently. And this might be, in an age during which the constant application of algorithmic procedures (such as intensive farming) to the biosphere might well cause biospheric collapse, what is known as a good thing.

Let's return to right versus left. Kant argues that knowing right from left, front from back, is how "time" and "space" are instantiated. So far, so embodied. At least one philosopher has found this intriguing, given how Kant is seen as the embodiment of disembodiment, someone whose habits one could set one's clock by as he went about his business in Königsberg.[13]

But what kind of body is Kant talking about? Why, the objectified body of an enslaved person or "animal." Terms such as "inch" or "foot" or "furlong" name nothing good. They name how many bodies of enslaved beings including nonhuman lifeforms can fit into a certain container or traverse a certain distance at a gallop suitable for delivering their rider or boss or owner to their destination as efficiently

[13] Edward Casey, "By Way of Body," in *The Fate of Place: A Philosophical History* (Berkeley: University of California Press, 1997), 202–242 (203–207).

as possible, to eliminate physical distance in order to achieve the projects of the subject or master. The desire of some Brexiteers to return to "imperial weights and measures" is exactly the desire of MAGA to "make great again" a country by establishing that "greatness" as one thing and one thing only: the way a white person measures their superiority to a Black person. Greatness is precisely the affect of the enslaver, of settler colonialism—the reason why America is so rich and powerful (its "greatness") having to do with the accumulation of a gigantic pile of wealth on the backs of enslaved humans.

Not to be able to tell right from left is thus to resist the "measures" that are indeed imperial insofar as they depend on a master–slave duality that underwrites the subject—object one—nay, every toxic duality remaining in existence, we claim: masculine–feminine, male–female, active–passive, revolutionary–incremental, essential–superficial. Furthermore, not being able to tell right from left does not mean being unable to "do" space and time. Rather it implies that one understands deeply, sometimes "only too well," what is good about Kant's thinking: the discovery of "feelings" that arise without conscious intent, "feelings" that drive thoughts and actions, feelings that are the very essence of embodiment and of elementality. This is a proprioception indeed, one that is not based on the master–slave duality and therefore one that is strictly speaking from the future, a time at which this duality has ceased to function as the template for all the others. We believe that this time is approaching. The template is snapping, becoming unsustainable in science, art, religion, economics, politics, group dynamics…

Which is the most optimistic explanation for the sudden upsurge in fascisms planet-wide. As group versions of what is called "the schizophrenic defense," they apply a frantic tourniquet to the gashing wound radical anxiety has dealt to the body of meaning. Morton's younger brother Steve suffers from schizophrenia, and one of his voices is fairly typical for paranoid schizophrenics insofar as it accuses him on a daily basis of being Adolf Hitler and for having caused the Holocaust.

But even this semantic hell is "easier" than floating in the void, easier than listening for the phenomenon to speak or prove its own meaning, easier than vibing. Being sucked into the black hole (albeit with a golden accretion disc) of Donald Trump provides meaning by eliminating the futural shimmer of meaningfulness in the form of plastic, readymade "facts" whose very value is that they are "alternative," to wit, they allow the fascist to plug holes without having to think, which would imply time, which would imply anxiety—which would imply feeling something. Better to be a droid empowered with the latest facial recognition software in the form of QAnon memes, and in possession of a gun, marching on the Capitol building utterly sure that one is the goodie. The fact that the goodie-in-chief is a criminal moron is a feature of this system, not a bug, in the same way that the Wagnerian Parsifal is Jesus minus the inconvenience of Jewish people and "the middle east" and non-whiteness.

Not to be able to tell right from left but instead to have a feeling-embodiment that is futural is most definitely a feature not a bug. It's a feature of a possible future in which the master–slave template has snapped. The fact that it is hard to imagine such a thing is why a specific kind of proprioception much closer to "vibing" or (non)meditation than calculating or mindfulness is so valuable. That difficulty exists precisely because it is "about" a lack of measurement. Ideas, remember, are not floating in a void: all ideas have an elemental and embodied quality; it's just that some of these qualities are not yet legitimated in social space.

And what if weakness as such were the superpower? Not just something identified as "a weakness," but weakness tout court? Weakness in itself? What if being weak and flawed was how to allow for genuine novelty in the world? It works in basic evolutionary terms and it works in spiritual terms. Morton's friend Ron Texada is a Christian who drives him to the airport sometimes as he runs a great taxi service. If you were to arrive in Morton's and Texada's hometown of Houston, Morton would think it compulsory that you take Ron's car, because Ron is a symptom of Houston at large. To travel in Ron's opulent car bedecked with golden toffee wrappers and to be driven around by Ron with his wraparound shades is to know Houston. Which is to say, you would enjoy what one might groaningly call "the bling of truth."

"Give it up to God," is what Ron says, whenever Morton complains about something. Whatever it is, "Give it up to God." What does it mean? In very simple terms—bracketing for a moment the existence or nonexistence of a deity, bracketing the specific qualities of that deity if they do exist—Ron means something quite precise. Ron means that a feeling is an idea from the future. Just letting the feeling be a feeling, to "park" it somewhere in one's experience, is the best way to cope with it.

In short, "Give it up to God" means, *be weak*. Weakness is the quintessential superpower in the same way in which wailing at the wall of *mathēsis* is how beauty happens. This is because weakness allows the future to be the future. Consider Peter's use in Acts 4:8–12 of the line in Psalms: "The stone that the builders rejected has become the cornerstone" (Psalm 118:22).[14] This has to do with the way in which there are very specific measurements for buildings such as temples, just as in Tibet, where there are very specific templates for thangka paintings of deities. Monks must paint them just so. A stone that does not accord with the template means one that cannot fit into the pre-established set of measurements for a building. To construct a building at all is most often to be in accord with a certain set of measurements, which embody certain sets of expectations about what buildings should be: in short, it is to embrace the past. The stone the builders rejected, then, is phenomenologically an object from the future. It is a mutant stone, as it were, and if there were a gathering of "stonist" stones, perhaps they would themselves reject this stone for not looking stony enough.

The weakness of the stone, its unfitness, is precisely what is most futural about it. The term "fit" implies measuring up to existing standards. This is why Darwin was leery of inserting the concept of fitness into *The Origin of Species*. In particular, the concept of fitness often implies a binary of sickness versus health, normal versus pathological. But what if the existing standards were also changing, because they belonged to "moving targets," themselves lifeforms subject to evolution?

[14] Bible, King James Version.

Darwin's initial analogy for evolution is as funny as it is deliberately inaccurate. Darwin essentially says, "You know pigeon fancying? How people breed pigeons to fly certain distances and have certain plumage? Now imagine all lifeforms doing that to all other lifeforms, but by accident."[15] Imagine, in other words, an uncountably infinite mesh of interconnected lifeforms, all "adapting to" one another. To adapt is also fishy. The mesh is immeasurably infinite because the number of evolutionary possibilities, how a certain mutation will pan out, is always uncertain; and because genetic mutation is random with respect to current need. "Adaptationism" is the name of a belief that mutations are "for" something, when they palpably are not.[16] When the nature documentary tells you that the duck's feet are "perfectly adapted" to swimming on the pond, the documentary is lying to you. Have you ever seen a coot? A coot does not have webbed feet. Coots exist. Evolution doesn't care about webbed feet.

What "fit" and "adapted" mean in Darwin is special. Darwin made these words as cheap as possible. The capitalist ideologue Herbert Spencer and the Christian Socialist Russell Wallace insisted that he put the terms in. "Survival of the fittest" was not Darwin's choice, and so Darwinism came *after* social Darwinism, because Darwin was nervous to publish his book. So forever Darwinism has been confused with social Darwinism, which is indeed about the survival of whoever has capitalist six-pack abs. Darwin cheapens the concept of "fittest," trims it down simply to mean that, if one is able to have children, they can keep whatever features one passes on. Dying before having children becomes the definition of unfit, which isn't that strict of a criterion. It leaves some wiggle room. If smartphone designers were in charge of evolution, ears would be tiny pinpricks because that would be more efficient. But evolution is not about efficiency.

Ultimately, then, evolution proceeds by mistakes and by the passing on of mutations in the "cheapest" possible way, so that a limit case for evolution would be a particularly weak lifeform whose weakness did not interfere with its ability to reproduce. A good example is a reptile, which is essentially a malfunctioning tadpole. Tadpoles can lose their tails; so can reptiles. Tadpoles can turn into frogs; reptiles cannot.

Reptiles are cold-blooded and require the sun's warmth to move. They therefore spend a lot of their time standing very still. This stillness has often provided a good model for human meditation, humans being mammalian in their search for prey and avoidance of predators, a dynamic that Buddhism refers to as samsara. Lizards accidentally take on the coloration of their surroundings, so that to say this is a "defense mechanism" isn't quite correct. Lizards escape being all eaten up because they just happen to be accidentally disguised. This accidental disguise then gets passed on to their offspring. Iguanas are forced to fish because they fall off rocks when a big wave comes along. To say that this is diving is never to have observed what is

[15] Charles Darwin, *The Origin of Species*, ed. Gillian Beer (Oxford and New York: Oxford University Press, 1996), 12, 18, 19–27.

[16] Richard Dawkins, *The Extended Phenotype: The Long Reach of the Gene* (Oxford and New York: Oxford University Press, 1999), 30.s.

happening to an iguana close up—and it *happens* to them, they don't *do* it. Free will is overrated?

Weakness is a fundamental superpower because what seems weak from the point of view of today's war will be strong tomorrow, metaphorically speaking. Pretty much every esoteric religious instruction manual planet-wide offers the dubious assurance that "The sun is always shining, even when there are clouds. Just rest without purpose or meaning, and the sun will shine on your third eye, the one you, as a lizard, have at the top of your head...or it won't."

Weakness as such has a phenomenology that is best called *silliness*. The word silly evokes a loving tenderness, as in Samuel Taylor Coleridge's poem *The Rime of the Ancient Mariner*

> :Oh sleep! it is a gentle thing,
> Beloved from pole to pole!
> To Mary Queen the praise be given!
> She sent the gentle sleep from Heaven,
> That slid into my soul.
>
> The silly buckets on the deck,
> That had so long remained,
> I dreamt that they were filled with dew;
> And when I awoke, it rained.[17]

The thing thought "silly" is in fact "worthy," something that is valued (Greek, *timao*, the etymon of Timothy). To be silly is to be blessed, and this in turn has the connotation of "weakness," as in the pejorative Texan "bless your heart," which Morton's grandfather unironically and unwittingly used as a simple and (for Morton) pleasant compliment. A driver who likes their car but is somewhat frustrated by it might choose to say "The blessed thing was stuck in first gear" instead of "The damned thing...". Or "Oh, bless!" is what a British person might say when a child says something naïve yet strangely on the money. We say "bless you" when one sneezes not simply because one is apparently ejecting an evil spirit but because sneezing is a way for the body to be spontaneous, that is to say, not measured in advance, not predictable: sneezing is silly, and silliness, as they say, is next to godliness. The second sense of "silly" in the *Oxford English Dictionary*, then, gets at the slide from holiness to weakness, "vulnerability, or physical incapacity," and thus to a sense of powerful "innocence," of a "helplessness" that condemns power.[18]

Silliness, as exemplified by the career of Jon Gnarr, founder of the Icelandic Best Party and erstwhile Mayor of Reykjavik, a position of considerable magnitude, ought to be a more frequent political tactic, and ought to replace "doubling down," the unedifying spectacle of someone's refusing to admit a mistake until they absolutely have to. "I don't know, I'll have to think about it and get back to you" is

[17] Samuel Taylor Coleridge, *The Rime of the Ancient Mariner*, in *Coleridge's Poetry and Prose*, ed. Nicholas Halmi, Paul Magnuson and Raimonda Modiano (New York: Norton, 2004), lines 292–300.

[18] *Oxford English Dictionary*, "silly," adj. n. and adv., A. I and II, accessed April 13, 2023, oed.com

exactly the wrong response to make to a journalist, which is of course why Gnarr (being a standup comedian unfazed by hecklers) used to say it during press conferences.[19] Its very weakness perfected the power of Gnarr's silliness.

The birth of Jesus was perhaps silly, too, insofar as it did not fit into pre-existing categories of thought. This was why the "Three Wise Men," an interdisciplinary team of scientists (Zoroastrian astrologers, in that time) sought out the baby born in the stable among "silly" nonhuman "animals." To this extent, Jesus became an anomalous scientific datum, indicating the necessity of new scientific facts, or of an entirely new paradigm. The common sight of a "crib scene" in a department store or church belies its fundamental silliness—in much the same way that any scientific data sufficiently unexpected according to the existing measuring technology is indistinguishable from miracle. (Which concept is in turn a decolonized version of the common idea that "any sufficiently powerful and strange technology is indistinguishable from magic"). Silly cows looking at a silly newborn is a potent subversion of the master–slave duality. That "vanilla" Christianities have such a power is unfashionable in many areas of academia. What is preferred is some kind of Gnosticism, even as the word itself is a pejorative term for "Christianities that Catholicism doesn't like" (just as "anarchist" is a pejorative term for "communism that Marxists don't like")—but Gnosticism is only chronologically prior in the history of Christianity. What came first was bound to be an anti-Semitic erasure of the Jewish body and a rewriting of the story as a Neoplatonic tale about a spiritual world trapped inside a physical one, which is—let's face it—just a religious version of the mind–body dualism that zealously participates in the master–slave duality. It is not a stretch to see how the Gnostic doctrine ("theory"?) that Jesus didn't really die on the Cross—that instead some kind of dummy Jesus died there to confuse people while the real spiritual Jesus resided safe in the spiritual beyond—provides a template for QAnon "theorists" to appropriate Gnosticism and overwhelm churchgoing thugs with signals far more potent than the doctrines their exasperated pastors in America are now failing to teach them.[20]

What is most radical about "official" Christianity is its doctrine of incarnation and resurrection, concepts that have precisely to do with silliness, with the weakest possible forms of embodiment and elementality. "King of Kings" does not mean "the head honcho, the highest possible Mafia boss beyond any of your puny conceptions of a boss." It means precisely "King in a loop," that is to say, the mathematical refutation of logical rules that forbid transfinite sets, the way a Jewish mathematician (Cantor) communed with God and the way a Catholic mathematician (Gödel) proved his value. (And here we note that the Greek word for timid, "*timao,*" also

[19] Jon Gnarr, private communication.

[20] Aaron Earls, "Half of U.S. Protestant Pastors Hear Conspiracy Theories in Their Churches," Lifeway Research, January 26, 2021, accessed April 14, 2023, https://research.lifeway.com/2021/01/26/half-of-u-s-protestant-pastors-hear-conspiracy-theories-in-their-churches/. Sophia Ankel, "Pastors Are Leaving Their Congregations after Losing Their Churchgoers to QAnon," *Business Insider*, March 14, 2021, accessed April 14, 2023, https://www.businessinsider.com/pastors-quit-after-qanon-radicalize-congregation-2021-3

carries the meaning "I esteem" or "I value," whether according to a preset price or according to an intrinsic value that exceeds and forestalls the possibility of presetting…).[21] Gnosticism is a feature of the master–slave template slipping off the slick surface of an entirely different way of feeling, a radical freedom of the broken body that rivals conventional ideas about the freedom of consciousness and renders "seeing" as "knowing," as pre-packaged power.

In this sense, "fear" is a feeling that we might re-examine, because the basic phenomenology of fear is not kowtowing to authority but *revering*. The former depends on the latter. In a sense, revering is fear in a loop—fear turned into the yoga pretzel of a Gödel number that makes it silly fear rather than sensible fear. Silliness and "coolness," a Yoruba philosophical concept that became a Black mode of resistance to the authority of segregation, align in the sense that they are futural, whereas "greatness" is the affect of "tradition" or, as Southerners put it with an oppressively knowing wink, "heritage."[22] America at its best makes things out of broken, that is to say malfunctioning, things, out of stones that the builders rejected: T-shirts are European clothes with bits cut off them. Likewise, the use of denim in jeans, diatonic scales in the blues, nonviolent tactics in protests (such as the badly named "dead weight" tactic of going limp during a protest of the murder of George Floyd in 2020), all these cool things are popular worldwide for a reason. America is a silly country that has not yet been born, because of slavery. America at its best could then be a template for whole new ways of embodying the elemental susurration of phenomena in this silly, contingent universe.

Just as Balds finds it hard to tell right from left, Morton stutters when he gets excited, and this is an index of "getting" something true in the strictly Kantian sense that beauty is truthfeel, but also in the strictly non-Kantian sense that truthfeel in the end is silly, not heavy. Truthfeel depends on letting oneself be vulnerable, on letting oneself stare calmly at a wall of seeming nonsense. In its practice resides the very definition of silliness, to wit, the humanities. For whoever controls the nonsense controls the world.

It remains only for the Bee Gees to give this essay its entirely inappropriate disco conclusion, lines whose falsetto embodiment is the quintessence of silliness in the service of truth and reconciliation. This silliness is thus in the service of a future world in which relations between lifeforms have taken on an "I and thou" format, not an instrumental one of "I and it," and this format is embodied in the vocal slide from a question that might imply an answer to the soft susurration of leaves in the wind, the quiet response to that call that simply repeats the question without expecting an answer, because the question

> is its own answer, because in the end, math is not calculation:How deep is your love?
> (How deep is your love, how deep is your love)
> I really mean to learn

[21] Eleuxis-web Greek lexicon, "timao," https://outils.biblissima.fr/fr/eulexis-web/?lemma=timao&dict=Bailly, accessed April 11, 2023.

[22] Robert Farris Thompson, *Flash of the Spirit: African and Afro-American Art and Philosophy* (New York: Vintage, 1984), 1–100 (6, 9, 12–21, 33, 37, 43, 47, 72–80).

Cause we're living in a world of fools
Breaking us down when they all should let us be
We belong to you and me.[23]

Bibliography

Ankel, Sophia. 2021, March 14. Pastors Are Leaving Their Congregations after Losing Their Churchgoers to QAnon. *Business Insider*. https://www.businessinsider.com/pastors-quit-after-qanon-radicalize-congregation-2021-3. Accessed 14 Apr 2023.

Bee, Gees. 1977. "How Deep is Your Love?" Recorded in 1977. In *On Saturday Night Fever: The Original Movie Sound Track*. RSO.

Bible. 1611. *King James Version*. London: Robert Barker.

Buber, Martin. 1971. *I and Thou*. Trans. Walter Kaufmann. New York: Touchstone.

Casey, Edward. 1997. By Way of Body. In *The Fate of Place: A Philosophical History, 202–242 (203–207)*. Berkeley: University of California Press.

Coleridge, Samuel Taylor. 2004. The Rime of the Ancient Mariner. In *Coleridge's Poetry and Prose*, ed. Nicholas Halmi, Paul Magnuson, and Raimonda Modiano, 292–300. New York: Norton.

Darwin, Charles. 1996. In *The Origin of Species*, ed. Gillian Beer, vol. 12, 18, 19–27. Oxford/New York: Oxford University Press.

Dawkins, Richard. 1999. *The Extended Phenotype: The Long Reach of the Gene, 30.s*. Oxford/New York: Oxford University Press.

Earls, Aaron. 2021, January 26. Half of U.S. Protestant Pastors Hear Conspiracy Theories in Their Churches. *Lifeway Research*. https://research.lifeway.com/2021/01/26/half-of-u-s-protestant-pastors-hear-conspiracy-theories-in-their-churches/. Accessed 14 Apr 2023.

Ferreira da Silva, Denise. 2007. *Toward a Global Idea of Race*. Minneapolis: University of Minnesota Press.

Harman, Graham. 2007. *Heidegger Explained: From Phenomenon to Thing*. Chicago: Open Court.

———. 2002. *Tool-Being: Heidegger and the Metaphysics of Objects*. Peru: Open Court.

Hart, Sarah. 2023, April 7. The Wondrous Connections Between Mathematics and Literature. *New York Times*. https://www.nytimes.com/2023/04/07/opinion/the-wondrous-connections-between-mathematics-and-literature.html?searchResultPosition=1. Accessed 14 Apr 2023.

Husserl, Edmund. 2008. Prolegomena to All Logic. In *Logical Investigations* (Trans. J.N. Findlay and Ed. Dermot Moran), vol 1, 1–161. London: Routledge.

James, Henry. 1908. Preface. In *The Spoils of Poynton*, ed. David Lodge and Patricia Quick, 126. New York: Heinemann.

Kant, Immanuel. 1999. *Critique of Pure Reason* (Translated and Edited by Paul Guyer), 143. Cambridge: Cambridge University Press.

Morton, Timothy. 2015. Elementality. In *Elemental Ecocriticism: Thinking with Earth, Air, Water and Fire*, ed. Jeffrey Cohen and Lowell Duckert. Minneapolis: University of Minnesota Press.

"'Silly,' adj. n. and adv." Oxford English Dictionary, A. I and II. Oed.com. Accessed 13 Apr 2023.

Smith, Douglas, Maurice Eggen, and Richard St. Andre. 2005. *A Transition to Advanced Mathematics*. 5th ed, 32. Boston: Brooks/Cole.

"Supplication to the Takpo Kagyü." 2011. Translated by Nalanda Translation Committee. Halifax: Vajradhatu.

Thompson, Robert Farris. 1984. *Flash of the Spirit: African and Afro-American Art and Philosophy*, 1–100 (6, 9, 12–21, 33, 37, 43, 47, 72–80). New York: Vintage.

[23] The Bee Gees, "How Deep Is Your Love?" (RSO, 1977), lyrics reproduced by permission; Martin Buber, *I and Thou*, tr. Walter Kaufmann (New York: Touchstone, 1971).

"Timao." Eleuxis-web Greek lexicon. https://outils.biblissima.fr/fr/eulexis-web/?lemma=timao&dict=Bailly. Accessed 11 Apr 2023.

Treena Balds completed her PhD in English at UC Davis in 2017 and has since been teaching 5th and 6th grade students in Montego Bay, Jamaica. She has published essays on Samuel Beckett and Mathematics and is currently exploring ways in which literature intersects with machine learning as well as with topology and complex analysis. She spends her very limited free time working on (dis)embodied fiction and purchasing video games she rarely gets to play.

Timothy Morton is Rita Shea Guffey Chair in English at Rice University and Director of the Cool America Foundation. Morton has collaborated with Laurie Anderson, Björk, Jennifer Walshe, Susan Kucera, Hrafnhildur Arnadottir, Sabrina Scott, Adam McKay, Jeff Bridges, Olafur Eliasson, Pharrell Williams and Justin Guariglia. Morton is the author of over twenty books, translated into seventeen languages, and of about three hundred essays. In 2014 Morton gave the Wellek Lectures in Theory at UC Irvine.

Chapter 7
A Biosemiotic Perspective on the Human Condition and the Environmental Crisis

Morten Tønnessen

Abstract This chapter presents a biosemiotic perspective on the basic situation for human beings and that of other organisms, with an emphasis on the subjective experience of sentient animals, and the sign use of all lifeforms. The human condition is portrayed as traditionally conceived, and then revisited in the new context of the current environmental crisis. A cornerstone of the text is an analysis of the materiality of the environmental crisis, and how the massive changes humans have caused in the physical environment can be understood in light of the semiotic agency of humans and other living beings. Experiential aspects of the environmental crisis are highlighted. The aim of the text is to improve our understanding of our species´ place in the natural world, our historical role in causing a global crisis for life, and how we can move forward towards a more sustainable future.

Keywords Environmental crisis · Human condition · Biosemiotics · Phenomenology · Semiotic agency · Semiotic causation · Anthropocentrism

Introduction

The environmental crisis reactualizes fundamental issues about what it means to be human. In the context of what is sometimes referred to as "the human condition," a crucial matter concerns how different, or similar, we are compared to other living beings. What can we learn from the different ways in which all organisms make use of signs? This chapter presents a biosemiotic perspective on the basic situation for human beings and that of other organisms, with an emphasis on sentient animals.

M. Tønnessen (✉)
Department of Social Studies, University of Stavanger, Stavanger, Norway
e-mail: morten.tonnessen@uis.no

L. Škof et al. (eds.), *Elemental-Embodied Thinking for a New Era*, Sophia Studies in Cross-cultural Philosophy of Traditions and Cultures 42,
https://doi.org/10.1007/978-3-031-42119-8_7

Applied to the context of the environmental crisis, a biosemiotic perspective can provide some of the framework that is required to properly understand our species′ place in the natural world, our historical role in causing a global crisis for life, and how we can move forward towards a more sustainable future.

Given that all organisms display forms of semiotic agency, a broad and inclusive view on phenomenology is warranted. Such a biosemiotic and eco-phenomenological perspective on the natural world is informative regarding understanding the human condition and the current environmental crisis, which is seldom analyzed from a broad experiential perspective implicating the subjective experiences of all sentient animals. By studying how semiotic causation supplements and interacts with efficient causation, we can shed light on how human beings are embodied in a material world and simultaneously strive to organize matter and ecology according to our human- and culture-specific agendas. It is a telling and disturbing fact that while the human body needs about 20 of the known chemical elements to function properly,[1] humankind today exploits absolutely all of the 81 non-volatile elements that constitute natural matter on Earth,[2] with little regard for the utility of these for other species. Arguably, current anthropocentrism and overexploitation can partly be explained by our lacking recognition of the agency, needs, and interests of other organisms, and a lacking understanding of our semiotic and material embodiedness in nature at large. As humans, we are naturally capable of symbolic and abstract thinking, but by disregarding more basic forms of sign exchange, and forgetting the impact that pursuing narrow self-interest has on the natural world, we lose contact with, and a sound sense of, the Earth that sustains us. A return to making use of our core experience as foundational for how we navigate in the world can help lead us in a more sustainable direction.

The Human Condition as Traditionally Conceived

The expression 'the human condition' lacks a precise, universally applied definition. It is often applied to matters of fundamental importance to human beings. While some use the expression to describe what will always apply to human beings, in a timeless fashion, others—such as Erich Fromm[3]—have used it to characterize the time they live in, with the challenges that are characteristic for humans of that period.

Those who refer to the human condition in a timeless fashion might think of it as something that is more or less synonymous with "human nature." However, whereas

[1] Maria Antonietta Zoroddu, Jan Aaseth, Guido Crisponi, Serenella Medici, Massimiliano Peana, and Valeria Marina Nurchid, "The Essential Metals for Humans: A brief Overview," *Journal of Inorganic Biochemistry* 195 (2019): 120.

[2] Roger-Maurice Bonnet and Lodewijk Woltjer, *Surviving 1000 Centuries: Can We Do It?* (Berlin-Heidelberg-New York/Chichester, UK: Springer/Praxis Publishing, 2008), 226, 242.

[3] Erich Fromm, "The Present Human Condition," *The American Scholar* 25, no. 1 (1955): 29–35.

talk of the human nature is common in moral and political discourse, "there are serious disagreements concerning the concept's content and explanatory significance—the starkest being whether the expression 'human nature' refers to anything at all."[4] Classical claims about what is uniquely human have portrayed human beings e.g., as 'political animals' (Aristotle) or 'rational animals' (Descartes). In modern times, such claims have been faced with 'the Darwinian challenge' from evolutionary theory, which has recontextualized human beings as natural beings on par with other organisms. Other advances in scientific knowledge about nonhumans have likewise required ever new rounds of critical reassessments of claims about allegedly unique human traits.

In recent decades, the expression 'The Human Condition' has been strongly associated with the political philosopher Hannah Arendt (1906–1975). In her book with the same name, she emphasizes that "the human condition is not the same as human nature, and the sum total of human activities and capabilities which correspond to the human condition does not constitute anything like human nature."[5] In Arendt's view, the human condition must be understood in light of the fact that human beings are conditioned beings, and the conditions we are met with could in principle change.

> The most radical change in the human condition we can imagine would be an emigration of men from the earth to some other planet. Such an event, no longer totally impossible, would imply that man would have to live under man-made conditions, radically different from those the earth offers him. [...] Yet even these hypothetical wanderers from the earth would still be human; but the only statement we could make regarding their "nature" is that they still are conditioned beings, even though their condition is now self-made to a considerable extent.[6]

As for whether or not the human condition is to be seen as changeable, Arendt takes the middle ground. What she offers is "a reconsideration of the human condition from the vantage point of our newest experiences and our most recent fears,"[7] treating "those general human capacities which grow out of the human condition and are permanent, that is, which cannot be irretrievably lost so long as the human condition itself is not changed."[8] Her book is said to deal "only with the most elementary articulations of the human condition, with those activities that traditionally, as well as according to current opinion, are within the range of every human being,"[9] and which "are fundamental because each corresponds to one of the basic conditions under which life on earth has been given to man."[10]

---

[4] Neil Roughley, "Human Nature," in *The Stanford Encyclopedia of Philosophy*, ed. Edward N. Zalta, Spring 2014.

[5] Hannah Arendt, *The Human Condition* (Chicago: The University of Chicago Press, 1958), 9–10.

[6] Ibid., 10.

[7] Ibid., 5.

[8] Ibid., 6.

[9] Ibid., 5.

[10] Ibid., 7.

Implicit in Arendt's narrative is a portrayal of a human tendency to have an ambivalent relation to the human condition. Arendt herself appears to share this ambivalence to some extent. In Arendt's view, "nothing entitles us to assume that man has a nature or essence in the same sense as other things."[11] While "[t]he earth is the very quintessence of the human condition, and earthly nature, for all we know, may be unique in the universe in providing human beings with a habitat in which they can move and breathe without effort and without artifice," Arendt stresses that the "human artifice of the world separates human existence from all mere animal environment."[12] In this sense, in her view, human reality is distinguished from the reality of any other living being on Earth. In a somewhat similar way, Erich Fromm holds that "Man's character has been molded by the demands of the world he has built with his own hands."[13]

Arendt describes "scientific endeavors [...] directed toward [...] cutting the last tie through which even man belongs among the children of nature"—in short, "the wish to escape the human condition."[14] Aspects of what she characterizes as "the rebellion against" the human condition, such as protests against the toil and trouble of labour, is "as old as recorded history"[15] —and have in recent centuries been supplemented by some philosophers´ dislike of bodily existence. In her own time, Arendt observes novel upheavals in the form of attempts to engineer a "future man, whom the scientists tell us they will produce in no more than a hundred years" and which implies "a rebellion against human existence as it has been given."[16] This latest rebellion carries with it a resistance against being restrained by customary Earthly conditions regarding phenomena such as natural births and death by disease, and means "that we, who are earth-bound creatures [...] have begun to act as though we were dwellers of the universe."[17]

Typically, when we claim that something relates to the human condition, we imply that it is something *all* humans, and *only* human beings, have to deal with. This might be warranted in some cases, but we should not rule out the possibility that some experiences or challenges that are of fundamental importance to human beings also apply to some cognitively complex animals. Whether or not this is the case is an important issue which must be examined in relation to the question of how unique human beings are.

As it happens, there is a strong correlation between assertions to the effect that human beings are so unique that we cannot even be compared to non-humans—i.e., that humans are incomparably unique—and the understanding that human beings

---

[11] Ibid., 10.

[12] Ibid., 2.

[13] Fromm, "The Present Human Condition," 29.

[14] Hannah Arendt, *The Human Condition, 2.*

[15] Ibid., 4.

[16] Ibid., 2.

[17] Ibid., 3.

are "alone in the universe" (or at the very least alone on Earth).[18] Existential loneliness of this sort correlates with human exceptionalism and is indicative of a lacking sense of community with non-human nature. The lack of identification with a larger natural community is in its turn often accompanied by a sense of alienation from nature, which is reinforced by the ways in which we arrange our daily lives in modern industrialized societies.

As Shaun Gallagher observes, "the concept of existential loneliness is said to be more fundamental [than other forms of loneliness], pervasive and part of the very structure of being human."[19] The concept's philosophical roots "are intertwined with the existential analyses of Heidegger and Sartre,"[20] both of whom regarded human beings as ontologically incomparable to other living beings. Overcoming this sense of existential loneliness and alienation from nature is part and parcel of the task of solving the environmental crisis.

One would perhaps think that human exceptionalism, if nothing else, unavoidably benefits human beings, but that is not necessarily the case. Srinivasan and Kasturirangan criticize what they call 'mainstream development,' which they hold to represent human exceptionalism, for entertaining "a very specific idea of human wellbeing: it envisages a 'good' human life as one that is freed from the vicissitudes—the risks and vulnerabilities—of living on the planet, of being a part of 'nature', of being animal."[21] Not only is human exceptionalism responsible for "rendering nonhuman life killable,"[22] giving the impression that "the instrumental use of other beings is acceptable in the pursuit of human wellbeing"[23]—it also frames human development as being

> about amplifying those human features that are believed to be maximally different from other species [...] and about pushing 'forward' those societies that do not meet these standards of development. Human ways of life that depart from the norms of human exceptionalism [...] are animalized and cast as in need of upliftment—of 'development.'"[24]

[18] Whether or not astrobiology with its hunch that "there must be somebody out there" can alleviate our longing for community with other intelligent beings like ourselves remains open to discussion. We feel superior among the species of Earth. If we, upon discovering more intelligent beings than ourselves elsewhere, were to acknowledge ontological inferiority rather than ontological superiority, it is conceivable that we might start experiencing a humbler type of loneliness.

[19] Shaun Gallagher, "A Critique of Existential Loneliness," *Topoi* (2023).

[20] Ibid.

[21] Krithika Srinivasan and Rajesh Kasturirangan, "Political Ecology, Development, and Human Exceptionalism," *Geoforum* 75 (2016): 126.

[22] Ibid.

[23] Ibid., 127.

[24] Ibid., 126.

Semiotic Agency in Humans and Non-humans

A foundational idea in biosemiotics is that all life makes use of, and relate to, signs. Among biosemioticians, a common understanding is that this implies that all organisms—as well as a number of other living systems—have semiotic agency.[25] Part of the basic situation for human beings and any other organism is that our relations to the environment, and to other living beings, are sign-mediated.

This biosemiotic perspective on the living world and on our place as human beings in the natural world has some fundamental implications for philosophical anthropology: First of all, in terms of sign use as such, human beings are *not* "alone in the universe." On the contrary, our capacity to understand and to make use of signs, and our predicament as a living being that has to relate to the world through signs, is principally something we share with everything that lives. Second, however, this does not at all imply that there is nothing unique about human beings and our sign use. The semiotic diversity that is to be found among lifeforms, and among species, is as significant as biological diversity by other measures. This calls for a more thorough characterization of the semiotic nature of human beings.

While human beings have *something* in common with all that lives, we have a lot more in common with (other) sentient animals, who, like us, are capable of reason, feeling, and having positive and negative experiences in the form of pleasure, pain, and the like. There are obvious physiological and behavioral similarities between us and them. Since experience and behaviour, as well as biological processes that occur within the body, are sign-based, we also have a lot in common in terms of specific sign processes.

One way to characterize what is unique about human beings in terms of sign use, is to refer to Charles Sanders Peirce's division of signs into symbols, icons, and indices.[26] While some semioticians hold that only human beings make use of symbols, I believe it is more accurate to state that human beings make *much more* use of symbols than any non-human does. An important point concerning symbols is that they are the most arbitrary class of signs. This has the effect that symbolic thought and imagery is rich, diverse, and amenable to change and innovation. In contrast, icons are characterized by their likeness to something else, and indices are characterized by some sort of relation to something else. While icons as well as indices tend to refer to something tangible in the natural world, symbols can pretty much refer to anything imaginable. The human capacity to make use of, understand, and invent symbols equips us with great and very flexible semiotic and cognitive resources. But all the advantages symbols have come with a risk of in effect decoupling our thinking and our experience from the natural world. In light of this, it can be said that the symbolic mode of thinking is simultaneously perhaps our greatest

[25] Alexei Sharov and Morten Tønnessen, *Semiotic Agency: Science beyond Mechanism* (Cham: Springer Nature, 2021).

[26] Charles Sanders Peirce, *Collected papers of Charles Sanders Peirce* (Cambridge, MA: Harvard University Press, 1931–1958).

human resource, and our Achilles heel as a species. This is because symbolic thought is a constitutive element of the varieties of anthropocentrism which inclines us to think, mistakenly, that 'everything is human.'

Our capacity for using symbols extensively and systematically enables us to apply language and abstract thinking. This has in turn resulted in the great cultural diversity found among humans, with thousands of cultures and languages giving rise to a considerable variation in how things are perceived by human beings. With this in mind, we should be cautious in our claims about what "the human way" of perceiving or acting amounts to.

There is no doubt, however, that the use of language is part of what distinguishes human capabilities from the semiotic capacities and behaviours of nonhumans. Here, too, there is considerable diversity. The phenomenologist David Abram addresses the differences in worldview found among oral and alphabetic cultures.[27] In this context, we should acknowledge that oral and written languages are equally symbolic—and yet be open to the idea that the symbolicity of language can play out very differently when transmitted in different media. Using a written language provides humans with incredible intellectual resources, for instance by enabling transfer of knowledge, information and narratives across time and space. But, as Abram narrates, transitioning from using an oral language to using a written language also tends to increase the distance to nature, as it were, in our felt relationship with the natural world. The apparently 'timeless' nature of written signs may, for one thing, have given rise to ideas about there being a dimension of reality which is equally timeless and just as real as the natural world we live in. This may have contributed to a devaluation of nature.

Above I have portrayed the semiotic agency of human beings *en masse*. It is worth pointing out, however, that if by agency we mean the capacity for effecting changes, then agency is not evenly distributed among human beings, but tends to reflect inequalities in income and consumption. The privileged in terms of consumption and currently high CO_2 emissions are also "those who are characterized by a high level of individual agency as well as having the organizational capabilities to actively exercise their agency."[28]

The Materiality of the Environmental Crisis

A proper understanding of the semiotic capabilities of human beings sheds new light on the materiality of the environmental crisis. In this section, I will start with describing aspects of the materiality of the environmental crisis in general, and then proceed to explain how this can be understood in a semiotic perspective. This

---

[27] David Abram, *The Spell of the Sensuous. Perception and Language in a More-Than-Human World* (New York: Vintage Books, 1997).

[28] Ilona M. Otto, Marc Wiedermann, Roger Cremades, Jonathan F. Donges, Cornelia Auer, and Wolfgang Lucht, "Human Agency in the Anthropocene," *Ecological Economics* (2020): 106463.

ultimately relies on an understanding of human ecology in which ecosystems can be understood in terms of sign flows, with special attention devoted to how human sign use affects ecosystems as a whole.

A first question is how living matter stands out from inanimate matter. In his description of the "universal nature of biochemistry," Norman R. Pace stresses that organisms as we know them require macromolecules, given that molecules that "serve terrestrial organisms typically are very large […] with molecular weights of thousands to millions of daltons"[29] (with one dalton being a unified atomic mass unit that is roughly equivalent to the weight of the lightest atom).

Life on Earth is often characterized as carbon-based. Of all the natural elements, carbon and silicon are unique in being "known to serve as the backbones of molecules sufficiently large to carry biological information," and while silicon "interacts with only a few other atoms," carbon "can readily engage in the formation of chemical bonds with many other atoms, thereby allowing for the chemical versatility required to conduct the reactions of biological metabolism and propagation."[30] A crucial feature of organic reactions, which involve carbon in interaction with other elements, is that they, unlike silicon-based reactions, are "broadly amenable to aqueous conditions."[31] This reflects the fact that water is the standard milieu for life, and typically accounts for a considerable part of the weight of living organisms.

Along with oxygen and hydrogen, carbon makes up more than 90% of the weight of a human body, with oxygen amounting for the most, followed by carbon and hydrogen. Oxygen is the most abundant element in most organisms, and also in the Earth's crust. In contrast, hydrogen is only the tenth most abundant element in the crust, and carbon only the 17th most abundant element, with an abundance of about 200 parts per million (ppm).[32] This illustrates how organisms stand out from matter in general in their material composition, and makes clear that organisms need to be systematically selective in their pursuit of maintaining their own materiality.

Altogether, around 60 elements are found in the human body, but only around 20 of these "are considered to be essential for life."[33] In humans as well as other organisms, deficiency symptoms indicate a lack of elements that are essential for life. To qualify as an essential element, its total absence from an organism must imply damage to vital functions.[34] Essential elements include essential metals. In a human organism, the medium amount of essential metals varies from about 1 kg of Calcium (Ca), which is mostly needed in the skeleton, to about 2 mg of Cobalt (Co), which is a key component of vitamin B12.[35] Iron (Fe) is "an essential element for

[29] Norman R. Pace, "The Universal Nature of Biochemistry," *PNAS* 98, no. 3 (2001): 805.

[30] Ibid.

[31] Ibid.

[32] "The most abundant elements in the Earth's crust," World Atlas, accessed May 28, 2023. https://www.worldatlas.com/articles/the-most-abundant-elements-in-the-earth-s-crust.html

[33] Zoroddu et al., "The Essential Metals for Humans: A Brief Overview," 127.

[34] Ibid., 121.

[35] Ibid.

practically all living systems" and occurs in a medium amount of about 5 g in a human body.[36]

This brief outline of how living matter stands out from inanimate matter underlines our dependence on, and intermingling with, the physical environment. A second question concerning the materiality of the environmental crisis is to what extent humans have altered physical environments on Earth. The fact that humankind is currently causing considerable changes in the physical environment is the key motivation behind designating our current geological epoch as 'the Anthropocene.'[37]

Jan Zalasiewicz and colleagues describe the 'physical technosphere' which has resulted from human activities, and which contains parts which are actively in use as well as material residue.[38] Material residue includes "gaseous components (e.g. carbon dioxide and methane that accumulate in the atmosphere)."[39] The biggest components of the technosphere in terms of mass are urban areas, rural housing, pastures, cropland, and trawled sea floor, with major components adding up to a mass of around 30 trillion Tons (Tt), "equivalent to > 50 kg/m^2 of the Earth's surface" or about 100.000 times as much as the weight of all humans.[40]

A third question is how human activities have affected biodiversity on Earth—whether it has happened by altering the physical environment, or by other means. Bar-On and colleagues estimate the distribution of current global biomass.[41] They also provide estimates of prehuman values for global biomass for terrestrial mammals compared to current values, which indicate the impact human civilization has had in the long run. These suggest that 100,000 BP, the biomass of wild mammals were about six times higher than today. On the other side, the current biomass of humans and livestock outdo the prehuman biomass of wild mammals by a factor of 4 or so[42] While wild mammals have been marginalized, then, the total biomass of mammals has increased quite substantially—but is now dominated by civilized (human) and captive (domesticated) mammals.

Combined, humans and livestock now account for an astonishing 96% of terrestrial mammal biomass, with humans alone accounting for 36%. In the bigger picture, however, mammals account for only a marginal share of animals, which in turn account for only a marginal share of global biomass (about 0.36%). At an estimated 550 gigatons of carbon (Gt C), global biomass is dominated by plants, followed by bacteria, fungi, Archaea, and protists—and only then animals.

---

[36] Ibid., 121, 124.

[37] Will Steffen et al., "The Anthropocene: Conceptual and Historical Perspectives," *Philosophical Transactions of the Royal Society A* 369 (2011): 842–867.

[38] Jan Zalasiewicz et al., "Scale and Diversity of the physical Technosphere: A geological Perspective," *The Anthropocene Review* 4, no. 1 (2016): 3.

[39] Ibid., 3–4.

[40] Ibid., 11.

[41] Yinon M. Bar-On, Rob Phillips, and Ron Milo, "The Biomass Distribution on Earth," *PNAS* 115, no. 25 (2018): 6506–6511.

[42] Ibid., Supplementary Information Appendix, 88 (Fig. S5).

Human civilization has had an impact far beyond mammals. Domesticated poultry now accounts for three times as much biomass as the global biomass of wild birds, and wild fish stocks have declined by an estimated 15% compared with their pre-human levels. While crops cultivated by humans amount for no more than about 2% of total plant biomass, human civilization "has also profoundly reshaped the total quantity of carbon sequestered by plants," leading to total plant biomass having "declined approximately twofold relative to its value before the start of human civilization."[43] In other words, there were about twice as much plant biomass in pre-human times.

As Burgess and Gaines observe, one of the interesting findings of the study of Bar-On and colleagues is that there is much more biomass on land than in the oceans—actually, by a factor of 80.[44] At the same time, "more than 70% of global animal biomass is found in the ocean."[45] In simplified terms, this allows us to say that "Earth has a plant-dominated landscape and an animal-dominated seascape."[46] Burgess and Gaines point out that in terrestrial and marine ecosystems alike, "the increasing human appropriation of global primary production has put disproportionate ecological pressure on other energy-intensive forms of life, namely large-bodied animals and top predators."[47] Overall, human appropriation of primary production is particularly intensive in cropland and infrastructure areas—and, in geographical terms, in Europe, Northern Africa, and Western and Southern Asia.[48]

The declines in biomass and biodiversity in different lifeforms can in many cases be mutually reinforcing and can be further challenged by climate change. For instance, about half of all plant species are dispersed by vertebrate animals.[49] The current "seed dispersal function has steeply declined from its natural level," and "defaunation has already limited the ability of animal-dispersed plants in many parts of the world to keep pace with climate change."[50]

A fourth and final question, which takes us back to the semiotic perspective applied in this chapter, is how human agency can have a so profound impact on ecosystems globally, despite our limited share of global biomass.

In this context, human impact on the environment can be outlined in terms of the expanding and eventually practically global range first of our own species and then, in our wake, also of several of our affiliated and favored species—notably livestock, and crop species—leading over time to the establishment of a global colonial

---

[43] Ibid.

[44] Matthew G. Burgess and Steven D. Gaines, "The Scale of Life and its Lessons for Humanity," *PNAS* 115 no. 25 (2018): 6328.

[45] Ibid.

[46] Ibid.

[47] Ibid., 6329.

[48] Helmut et al. Haberl, "Quantifying and Mapping the Human Appropriation of Net Primary Production in Earth's Terrestrial Ecosystems," *PNAS* 104, no. 31 (2007): 12944.

[49] Evan C. Fricke et al., "The Effects of Defaunation on Plants' Capacity to Track Climate Change," *Science* 375 no. 6577 (2022): 210.

[50] Ibid., 212, 213.

organism.[51] The implied ecological empire, which is hierarchically organised with *Homo sapiens* on top, has "provided global breeding grounds" for some species, while leaving others marginalized.[52]

In a biosemiotic perspective, the notion of semiotic agency is crucial for understanding human agency. In his Ph.D. thesis, cognitive semiotician Juan Carlos Mendoza-Collazos introduces the notion of 'enhanced agency,' which is a useful supplement, stating that the notion "implies the prosthetic incorporation of artefacts into the agentive capabilities of the agents."[53] In his view, all humans are endowed with enhanced agency, which expands human beings´ capacity for action.

Part of the explanation of humans´ immense impact, then, is related to the human use of domesticated animals and plants, and of tools and artefacts, which expands our species´ semiotic agency and ecological reach. The biosemiotician Jesper Hoffmeyer, who acknowledges "agency as an inherent property of living systems,"[54] introduces the notion of 'semiotic causation' or 'semiotic causality,' which he defines as "*bringing about things under guidance of interpretation in a local context*,"[55] or "*bringing about effects through interpretation*."[56] In Hoffmeyer's conception, semiotic causation "must always operate through the mechanisms of material efficient causation."[57] It "cannot be reduced to efficient causality, but is dependent on efficient causality since interpretative activity, even in its most primitive modes, is connected to possible anticipatory action, and action unquestionably depends on efficient causality."[58] Hoffmeyer sums up his view by stating that semiotic causality "thus gives direction to efficient causality, while efficient causality gives power to semiotic causality."[59]

In a biosemiotic perspective in the tradition of Hoffmeyer, all organisms are capable of interpretation, and therefore capable of triggering semiotic causation. Figure 7.1 shows the interplay between semiotic and efficient causation for an organism endowed with an Umwelt in its dealings with the physical environment (*Umgebung*). As illustrated by the figure, the Umwelt, or subjective, experienced world of the organism in a broad sense can more precisely be divided into an

---

[51] Morten Tønnessen, "The Global Species," *New formations: a journal of culture/theory/politics* 69 (2010): 98.

[52] Ibid.

[53] Juan Carlos Mendoza-Collazos, *Agency and Artefacts: A Cognitive Semiotic Exploration of Design* (Lund: Lund University, 2022): 24.

[54] Jesper Hoffmeyer, "Why do we need a Semiotic Understanding of Life?", in *Beyond Mechanism. Putting Life Back into Biology*, ed. B.G. Henning and A.C. Scarfe Janham (Boulder, New York, Toronto, Plymouth, UK: Lexington Books, 2013), 157.

[55] Ibid., 158.

[56] Jesper Hoffmeyer, "Semiotic Scaffolding of Living Systems," in *Introduction to Biosemiotics: The New Biological Synthesis*, ed. Marcello Barbieri (Dordrecht: Springer, 2008), 152.

[57] Ibid.

[58] Jesper Hoffmeyer, "Semiotic Individuation and Ernst Cassirer's Challenge", *Progress in Biophysics and Molecular Biology* 119 (2015): 610.

[59] Ibid.

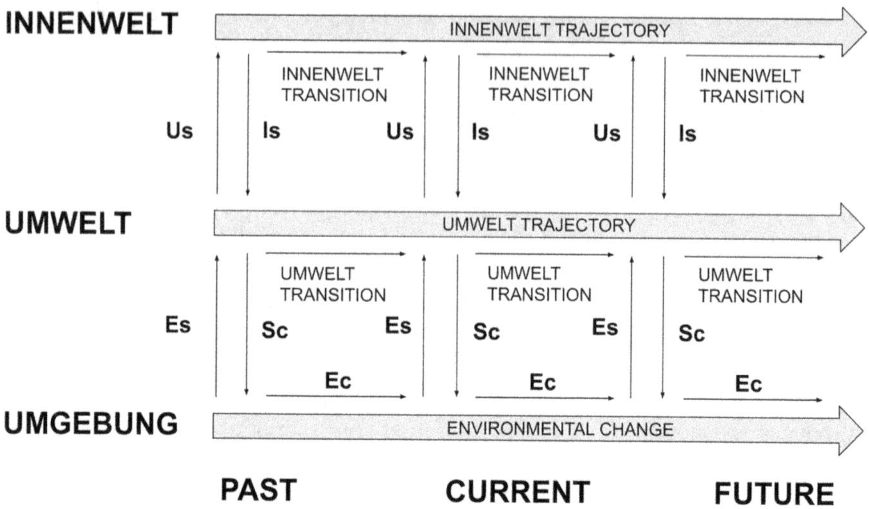

Fig. 7.1 Three-dimensional interactive semiotic model of environmental change. Abbreviations: *Ec* Efficient causation, *Es* Environmental signals, *Is* Innenwelt signals, *Sc* Semiotic causation, *Us* Umwelt signals. (From Morten Tønnessen 2019: 420)

Innenwelt (inner subjective world, within the organism) and an *Umwelt* (outer subjective world, surrounding the organism).[60]

In principle, all organisms are capable of both impacting the physical environment they live in and adjusting to it. The *Umwelt transitions* organisms endowed with an Umwelt go through are of different kinds, some are regular, while others constitute historical events.[61] The environmental crisis involves several of the latter kind, and as a global crisis, it can be conceived of as a massive global Umwelt transition.

In light of semiotic causation, we should acknowledge that all organisms, as sign users, are directly affected by the environmental crisis, and that their lives change as the sign flows of ecosystems are affected by human actions.

Experiential Aspects of the Environmental Crisis

Using the outlook of biosemiotics as a starting point, it makes sense to emphasize experiential aspects of the environmental crisis, with a scope covering the subjective experience of human beings as well as the subjective experience of all sentient

[60] Jakob von Uexküll, *Umwelt und Innenwelt der Tiere* (2nd ed.) (Berlin: Verlag von Julius Springer, 1921).

[61] Morten Tønnessen, "Umwelt Transitions: Uexküll and Environmental Change," *Biosemiotics* 2, no. 1 (2009): 47–64.

animals. The latter—animal experience—is seldom granted much attention in mainstream science and policymaking. For some reason, the animals´ experience is acknowledged, to some extent, in the context of animal welfare, with its focus on individuals of animals, but typically disregarded in the context of environmental problems, where a more systemic focus is common. While this might follow from the different framing of animals in different policy contexts, the different framing in itself cannot legitimately justify the disregard for subjective experience in the context of environmental problems.

Analyzing the environmental crisis from the vantage point of subjective, sentient experience has implications for how we conceive of the environmental crisis. If subjective experience is emphasized, then the ways in which environmental problems affect the lives of individual human beings and sentient animals must be emphasized. This will tend to favorize views on the environmental crisis that are perceptive with regard to damages and threats to living conditions and wellbeing. The inclusion of *animal* experience points towards a more equitable view on the interests of humans and animals, which conflicts with human exceptionalism with its one-sided prioritization of human interests only.

In the case of human perception of nature, it is important to be aware that our specifically human Umwelt is layered, as it were, with different kinds of experience involved. Distinctions can be made between our *core* Umwelt experience, our *mediated* Umwelt experience, and our *conceptualized* Umwelt experience.[62] Only the first kind of experience involves direct encounters with other living beings or with nature. Often when we relate to nature, we do so in an indirect way, mediated by ideas, imagery, or the like, rather than by way of our own encounters. In our symbol-laden modern cultures, many of our perceptions—and beliefs—are shaped in a sphere where there is little room for tangible nature experiences.

The perspective of biosemiotics calls for a broad and inclusive view on phenomenology which acknowledges phenomena beyond human phenomena. Contemporary endeavors to naturalize phenomenology are supported by biosemiotics and should be extended to include a reconceptualization of natural history. A biosemiotic narrative on natural history ultimately amounts to portraying the natural history of the phenomenal world, with its stepwise emergence of new layers of semiotic, cognitive, and phenomenal complexity, interrupted by periodical crises. A chapter in this story concerns the current environmental crisis. An eco-phenomenological re-telling of human—nature relations and ecological developments leading up to our current crisis is highly relevant in helping us to understand the Anthropocene as a geological era dominated by human agency.[63]

When considering the experiential aspects of the environmental crisis, we should bear in mind how global biomass and biodiversity is distributed, as discussed earlier in this chapter. Many would be inclined to acknowledge subjective experience in a proper sense only in sentient animals. That leaves out the organismic circumstances

[62] Morten Tønnessen, "Umwelt Trajectories," *Semiotica* 198 (2014): 159–180.

[63] Will Steffen et al., "The Anthropocene: Conceptual and Historical Perspectives."

of all plants, bacteria, fungi, Archaea, protists, and viruses, which taken together amount for the vast majority of individual organisms in nature, and a dominant share of global biomass.[64] In a biosemiotic perspective, however, all these lifeforms also make use of signs, and through their sign use, they are directly affected by the ways in which humans have altered their environments.

As for animals, we should recall that the majority of animal species are insect species, and that most animals in terms of biomass live in the oceans.[65] Despite the human dominance of terrestrial mammals, the typical animal is either a miniscule insect—or a marine arthropod. And the most typical animal environment is arguably not terrestrial, but marine and aqueous.[66]

The Human Condition Reconsidered

Some scholars have argued that the human condition is fundamentally changed given our current predicament involving the environmental crisis. Dipesh Chakrabarty refers to a "collapsing of multiple chronologies—of species history and geological times into our very own lifetimes, within living memory,"[67] and concludes that

> the human condition has changed. This changed condition does not mean that the related but different stories of humans as a divided humanity, as a species, and as a geological agent have all fused into one big story, and a single story of the planet and of the history of life on it can now serve in the place of humanist history. As humans we have no way of experiencing […] these other modes of being that are also open to us today. Humans, humans as a species, and humans as the makers of the Anthropocene […] are agents of very different kinds.[68]

While Hannah Arendt did not relate to the environmental crisis as such, she did acknowledge in her classical work on the human condition that "there is no reason to doubt our present ability to destroy all organic life on earth."[69] In a reading of Arendt, Anna Yeatman notes that the "human way of being alive both ties humans to other creaturely beings and differentiates humans from them. But being alive, as such, is what these beings have in common, and it indicates a shared dependence on an earth-bound existence."[70] A biosemiotic understanding of the basic situation for

---

[64] Bar-On et al., "The Biomass Distribution on Earth."

[65] Ibid.

[66] Burgess and Gaines, "The Scale of Life and its Lessons for Humanity," 6328.

[67] Dipesh Chakrabarty, "The Human Condition in the Anthropocene," in *The Tanner Lectures in Human Values, Vol. 35*, ed. Mark Matheson (Salt Lake City: University of Utah Press, 2016), 180.

[68] Ibid.

[69] Arendt, *The Human Condition*, 3.

[70] Anna Yeatman, "The Human Condition in the Anthropocene," in *Manifesto for Living in the Anthropocene*, ed. Katherine Gibson, Deborah Bird Rose, and Ruth Fincher (Brooklyn, NY: Punctum books, 2015), 124.

humans and nonhumans alike as fellow sign users takes this outlook an important step further.

Masatake Shinohara proposes, as part of a critique of Arendt, that "we may consider the contemporary ecological crisis as causing the breakdown of the human world and its collision with terrestrial reality."[71] He stresses that "what conditions human beings in the most fundamental sense includes not only the world of the human artifact but also the world of earthly things," which is ultimately "beyond any human measurement."[72] In his view, "the consideration of the human condition should be fundamentally reformulated" by making the human world "open to the earthly things that vastly expand outside of the human artifice," and he suggests that "the existential condition upon which human livings depends might be reconstructed within the vastness of the earthly reality, becoming part of it."[73] This appears to be compatible with a biosemiotic understanding of the basic situation for humans and other living beings.

Paths Towards Sustainable Futures

Coming from a background as an economist, John Maynard Keynes claimed, almost a hundred years ago, that "the economic problem, the struggle for subsistence, always has been hitherto the primary, most pressing problem of the human race"[74]—and "not only of the human race, but of the whole of the biological kingdom from the beginnings of life in its most primitive forms."[75] In his view, "we have been expressly evolved by nature—with all our impulses and deepest instincts—for the purpose of solving the economic problem."[76]

Keynes envisioned that the economic problem, the struggle for subsistence, would be solved around our current times, and that this would change the living conditions for humans fundamentally—if only old habits would allow us. He did not "see the solution to the economic problem as limited by human nature (unlimited wants)"—in this sense Keynes was optimistic, and foresaw a post-scarcity society.[77]

For a majority of humankind, though not yet for all, this has now come true in terms of material affluence. But at what cost? Given the ecological crisis, we have

---

[71] Masatake Shinohara, "Rethinking the Human Condition in the Ecological Collapse," *The New Centennial Review* 20, no. 2 (2020): 179.

[72] Ibid., 180.

[73] Ibid., 195.

[74] John Maynard Keynes, "Economic Possibilities for our Grandchildren," in *Essays in Persuasion* (New York: W.W. Norton & Co., 1963).

[75] Ibid.

[76] Ibid.

[77] Robert Chernomas, "Keynes on Post-Scarcity Society," *Journal of Economic Issues* 18, no. 4 (1984): 1009.

not yet arrived at a sustainable society. A truly sustainable society will have to be sustainable both in social terms and in environmental terms. This will have to include reducing resource use overall. Unfortunately, it has been demonstrated that the United Nations´ Sustainable Development Goals overall favors further economic growth, rather than resource containment.[78] In the context of energy needs, we should consider how much of Earth's biomass humans are already exploiting. An anticipated further increase in biomass harvest for 'green' bioenergy purposes "would almost double the present biomass harvest and generate substantial additional pressure on ecosystems."[79]

Some of the foundational ideas of ecological economics are still very pertinent under current circumstances. As Herman Daly stipulated already in 1968, "the ultimate subject matter of biology and economics is one, viz., the life process."[80] This perspective requires us to reconceptualize the economy from being conceived of as a solely human enterprise to being considered as a more comprehensive natural economy which also encompasses the "non-human economy"[81] and all "the exchanges of economic and ecological commodities making up the total economy of life."[82] While some progress has been made in this respect—ecological economics is now as blossoming field of study—we are still far from a truly ecological understanding of economics in the mainstream. To make further progress, mainstream economists and policy makers must wrap their heads around the idea that human beings are not the only stakeholders in the context of the economy, that there are resources that matter to nun-humans, too, and that there are values and valuers beyond human ones.[83] In short—in economics, as well as in other fields of political relevance, human exceptionalism must be overcome, and the agency of nonhumans must be acknowledged.

As mentioned earlier, the human Umwelt is layered, and involves mediated and conceptual aspects in addition to core Umwelt experience related to our direct encounters with other living beings and with nature. With our species-specific preference for symbolicity, we run a risk of disassociating from tangible nature experiences and being alienated from nature. As a countermeasure against such tendencies, we need to cultivate a culture of appreciation for more basic forms of sign exchange which is of fundamental importance to ecological functioning. This concerns how

---

[78] Nina Eisenmenger et al., "The Sustainable Development Goals prioritize Economic Growth over Sustainable Resource Use: A Critical Reflection on the SDGs from a Socio-Ecological Perspective," *Sustainability Science* 15 (2020): 1101–1110.

[79] Haberl et al., "Quantifying and Mapping the Human Appropriation of Net Primary Production in Earth's Terrestrial Ecosystems," 12946.

[80] Herman E. Daly, "On Economics as a Life Science," *Journal of Political Economy* 76, no. 3 (1968): 392.

[81] Ibid., 401.

[82] Ibid., 403.

[83] Morten Tønnessen, "The True Value of 'Doing Well' Economically," in *Innovation and the Arts: The Value of Humanities Studies for Business*, ed. Piero Formica and John Edmondson (Bingley: Emerald Publishing, 2020): 91–109.

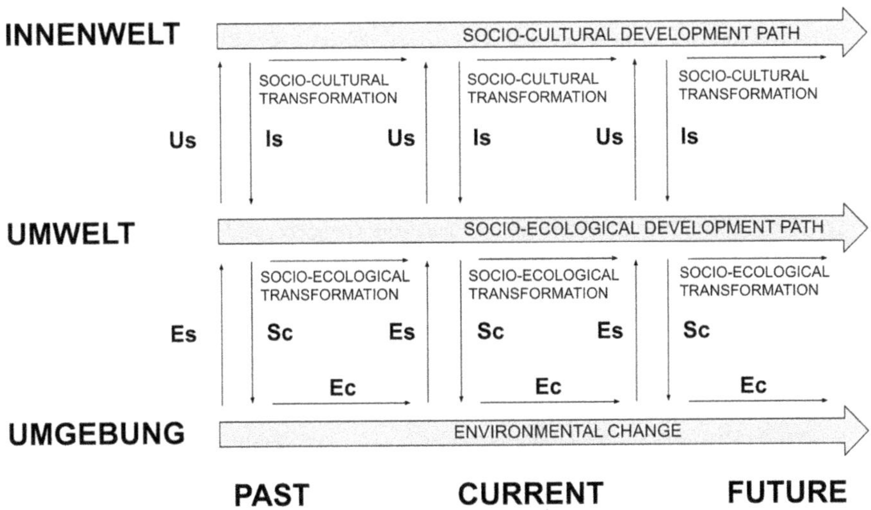

Fig. 7.2 Three-dimensional interactive semiotic model of societal transformations. Abbreviations: *Ec* Efficient causation, *Es* Environmental signals, *Is* Innenwelt signals, *Sc* Semiotic causation, *Us* Umwelt signals. (From Morten Tønnessen, "Anticipating the Societal Transformation required to Solve the Environmental Crisis in the 21st Century," 26)

we look at nature beyond human nature, and also how we choose to spend our time, engaging with nonhuman nature.

Implicit in the change that is needed to overcome the environmental crisis is a growing sense of acknowledgement of the varieties of nonhuman subjective experience. To overcome anthropocentrism and human exceptionalism, we must start by overcoming our habitual obsession with narrow self-interest. A further step in the right direction would involve adapting more to the seasons, rhythms, and local variations of nature, rather than trying to engineer nature so as to match human standards and expectations.

The changes that are needed to embark upon paths toward sustainable futures can also be understood in the perspective of a reworked version of the three-dimensional interactive semiotic model of environmental change presented in Fig. 7.1. Figure 7.2 shows a similar model applied to human societies. Changes both in the human Innenwelt (values, beliefs, identity etc.) and the human Umwelt (perceptions, behaviors, etc.) are required to achieve socio-cultural and socio-ecological transformations towards sustainable futures.

Bibliography

Abram, David. 1997. *The Spell of the Sensuous. Perception and Language in a More-Than-Human World*. New York: Vintage Books.

Arendt, Hannah. 1958. *The Human Condition*. Chicago: The University of Chicago Press.

Bar-On, Yinon M., Rob Phillips, and Ron Milo. 2018. The Biomass Distribution on Earth. *Proceedings of the National Academy of Sciences (PNAS)* 115 (25): 6506–6511. https://doi.org/10.1073/pnas.1711842115. Includes "Supplementary Information Appendix" (113 pp.).

Bonnet, Roger-Maurice, and Lodewijk Woltjer. 2008. *Surviving 1,000 Centuries: Can We Do It?* Berlin/Heidelberg/New York/Chichester: Springer/Praxis Publishing.

Burgess, Matthew G., and Steven D. Gaines. 2018. The Scale of Life And its Lessons for Humanity. *Proceedings of the National Academy of Sciences (PNAS)* 115 (25): 6328–6330. https://doi.org/10.1073/pnas.180701911.

Chakrabarty, Dipesh. 2016. The Human Condition in the Anthropocene. In *The Tanner Lectures in Human Values*, ed. Mark Matheson, vol. 35, 137–188. Salt Lake City: University of Utah Press.

Chernomas, Robert. 1984. Keynes on Post-Scarcity Society. *Journal of Economic Issues* 18 (4): 1007–1026.

Daly, Herman E. 1968. On Economics as a Life Science. *Journal of Political Economy* 76 (3): 392–406.

Eisenmenger, Nina, et al. 2020. The Sustainable Development Goals prioritize Economic Growth over Sustainable Resource Use: A Critical Reflection on the SDGs from a Socio-Ecological Perspective. *Sustainability Science* 15: 1101–1110. https://doi.org/10.1007/s11625-020-00813-x.

Fricke, Evan C., Alejandro Ordonez, Haldre S. Rogers, and Jens-Christian Svenning. 2022. The Effects of Defaunation on Plants' Capacity to Track Climate Change. *Science* 375 (6577): 210–214. https://doi.org/10.1126/science.abk3510.

Fromm, Erich. 1955. The Present Human Condition. *The American Scholar* 25 (1): 29–35.

Gallagher, Shaun. 2023. A Critique of Existential Loneliness. *Topoi*. https://doi.org/10.1007/s11245-023-09896-4.

Haberl, Helmut, et al. 2007. Quantifying and Mapping the Human Appropriation of Net Primary Production in Earth's Terrestrial Ecosystems. *Proceedings of the National Academy of Sciences (PNAS)* 104 (31): 12942–12947. www.pnas.org/cgi/doi/10.1073/pnas.0704243104.

Hoffmeyer, Jesper. 2008. Semiotic Scaffolding of Living Systems. In *Introduction to Biosemiotics: The New Biological Synthesis*, ed. Marcello Barbieri, 149–166. Dordrecht: Springer.

———. 2013. Why Do We Need a Semiotic Understanding of Life? In *Beyond Mechanism. Putting Life Back into Biology*, ed. B.G. Henning and A.C. Scarfe Janham, 147–168. Boulder/New York/Toronto/Plymouth: Lexington Books.

———. 2015. Semiotic Individuation and Ernst Cassirer's Challenge. *Progress in Biophysics and Molecular Biology* 119: 607–615.

Keynes, John Maynard. 1963. Economic Possibilities for our Grandchildren. In *Essays in Persuasion*, 358–373. New York: W.W. Norton & Co.. (first published 1930).

Maran, Timo. 2014. Place and Sign. Locality as a Foundational Concept for Ecosemiotics. In *Re-imagining Nature: Environmental Humanities and Ecosemiotics*, ed. A.K. Siewers, 79–89. Lewisburg: Bucknell University Press.

Mendoza-Collazos, Juan Carlos. 2022. *Agency and Artefacts: A Cognitive Semiotic Exploration of Design*. Doctoral dissertation. Lund: Lund University.

Otto, Ilona M., Marc Wiedermann, Roger Cremades, Jonathan F. Donges, Cornelia Auer, and Wolfgang Lucht. 2020. Human Agency in the Anthropocene. *Ecological Economics*: 106463. https://doi.org/10.1016/j.ecolecon.2019.106463.

Pace, Norman R. 2001. The Universal Nature of Biochemistry. *Proceedings of the National Academy of Sciences (PNAS)* 98 (3): 805–808.

Peirce, Charles Sanders. 1931–1958. *Collected Papers of Charles Sanders Peirce*. Edited by Charles Hartshorne and Paul Weiss (vol. 1–6) and Arthur W. Burks (vol. 7–8). Cambridge, MA: Harvard University Press.

Roughley, Neil. Human Nature. In *The Stanford Encyclopedia of Philosophy* (Spring 2021 Edition), ed. Edward N. Zalta. https://plato.stanford.edu/archives/spr2021/entries/human-nature/.

Sharov, Alexei, and Morten Tønnessen. 2021. *Semiotic Agency: Science Beyond Mechanism*, Biosemiotics 25. Cham: Springer Nature.

Shinohara, Masatake. 2020. Rethinking the Human Condition in the Ecological Collapse. *The New Centennial Review* 20 (2): 177–203.
Srinivasan, Krithika, and Rajesh Kasturirangan. 2016. Political Ecology, Development, and Human Exceptionalism. *Geoforum* 75: 125–128.
Steffen, Will, Jacques Grinevald, Paul Crutzen, and John McNeill. 2011. The Anthropocene: Conceptual and Historical Perspectives. *Philosophical Transactions of the Royal Society A* 369: 842–867. https://doi.org/10.1098/rsta.2010.0327.
Tønnessen, Morten. 2009. Umwelt Transitions: Uexküll and Environmental Change. *Biosemiotics* 2 (1): 47–64. https://doi.org/10.1007/s12304-008-9036-y.
———. 2010. The Global Species. *New Formations: A Journal of Culture/Theory/Politics* 69: 98–110.
———. 2014. Umwelt Trajectories. *Semiotica* 198: 159–180. https://doi.org/10.1515/sem-2013-0106.
———. 2019. What Can Be Known About Future Umwelten? *The American Journal of Semiotics* 35 (3–4): 401–429. https://doi.org/10.5840/ajs202012359.
———. 2020. The True Value of 'Doing Well' Economically. In *Innovation and the Arts: The Value of Humanities Studies for Business*, ed. Piero Formica and John Edmondson, 91–109. Bingley: Emerald Publishing. https://doi.org/10.1108/978-1-78973-885-820201005.
———. 2021. Anticipating the Societal Transformation required to Solve the Environmental Crisis in the 21st Century. *Sign Systems Studies* 49 (1/2): 12–62. https://doi.org/10.12697/SSS.2021.49.1-2.02.
Von Uexküll, Jakob. 1921. *Umwelt und Innenwelt der Tiere*. 2nd ed. Berlin: Verlag von Julius Springer.
Yeatman, Anna. 2015. The Human Condition in the Anthropocene. In *Manifesto for Living in the Anthropocene*, ed. Katherine Gibson, Deborah Bird Rose, and Ruth Fincher, 123–126. Brooklyn: Punctum books.
Zalasiewicz, Jan, et al. 2016. Scale and Diversity of the Physical Technosphere: A Geological Perspective. *The Anthropocene Review* 4 (1): 1–14.
Zoroddu, Maria Antonietta, Jan Aaseth, Guido Crisponi, Serenella Medici, Massimiliano Peana, and Valeria Marina Nurchid. 2019. The Essential Metals for Humans: A Brief Overview. *Journal of Inorganic Biochemistry* 195: 120–129.

Morten Tønnessen (born 1976) is Professor of philosophy at University of Stavanger's Department of social studies, Stavanger, Norway. His Ph.D. thesis, "Umwelt transition and Uexküllian phenomenology: An ecosemiotic analysis of Norwegian wolf management," was defended at University of Tartu (Estonia) in 2011. He served as an Editor-in-Chief of *Biosemiotics* 2013–2020, and has been Secretary (2011–2017 and 2023–) and President of the Nordic Association for Semiotic Studies (2017–2023). Recent publications include *Semiotic agency: Science beyond mechanism* (with Alexei Sharov, Springer 2021) and "Anticipating the societal transformation required to solve the environmental crisis in the 21st century" (in *Sign Systems Studies*, 2021).

Chapter 8
The Gravity of Our Situation: On Acrophobia and Eco-Paralysis

Ole Martin Sandberg

Abstract This chapter takes gravity as a fundamental element. Gravity is one of the physical forces without which there would not be life nor matter as we know it. Gravity can keep us grounded, but it can also be terrifying and discomforting. As a person suffering from acrophobia, I give a phenomenological account of how this condition is not merely a "fear of heights" but a debilitating somatic condition that affects the sensory apparatus and my being in the world. Acrophobia is one of many environment and place related phobias, which disrupt the harmonious relation between environment and body. I then relate this experience to the environmental and social challenge of climate change—a problem that is so immense it can give a similar feeling of debilitating dizziness. "Eco-paralysis" is different from "climate anxiety" just as acrophobia is different from a mere "fear of heights." It is an embodied inability to act when faced with the complexity and gravity of our situation: One has to look away from the problem in order to regain sensory control. The challenge of our time is to find the balance, which involves focusing our attention on the ground while still being aware of the environmental abyss.

Keywords Acrophobia · Climate anxiety · Eco-Paralysis · Environmental experience · Phenomenology · Gravity

This chapter takes gravity as a fundamental element. Gravity is one of the physical forces without which there would not be life nor matter as we know it, but it is also essential for our embodied being in the world. It holds us to the planet, structures our movement in the world and our relations to it. Gravity is not only an elemental ontological force of the world, it is also a force that has conditioned the evolutionary and individual development of the human body and our sensory apparatus. It is thus an elementary epistemic force that shapes our thinking and phenomenological

O. M. Sandberg (✉)
Institute of Philosophy, University of Iceland, Reykjavik, Iceland
e-mail: oms@hi.is

L. Škof et al. (eds.), *Elemental-Embodied Thinking for a New Era*, Sophia Studies in Cross-cultural Philosophy of Traditions and Cultures 42,
https://doi.org/10.1007/978-3-031-42119-8_8

dispositions. Gravity not only ties us to the Earth in the literal and physical sense; it also makes us Earthly beings. Gravity can thus keep us grounded, but it can also be terrifying and discomforting. As a person suffering from acrophobia, I give a phenomenological account of how this condition is not merely a "fear of heights" but a debilitating somatic condition that affects the sensory apparatus and my being in the world: when it is triggered, my experience of the world shrinks, my body is paralyzed, and I am unable to focus on the task at hand. Conditions like this, and other environment and place related phobias, problematize the more romantic views of nature experience which focus on harmony between the body and environment. Different environments can cause various kinds of discomfort for different people.

I then relate this experience to the environmental and social challenge of climate change. The experience of acrophobia is but a more localized instance of the relation many of us have to the current state of the world—i.e., regarding the multiple and interwoven crises of global warming, mass extinction, pollution, and the possibility of the collapse of the foundations of human existence. These environmental crises are so massive that they are hard, if not impossible, to comprehend. The philosopher Timothy Morton calls this type of thing "hyperobjects."[1] These hyperobjects are like the abyss that triggers my acrophobia: they induce dizziness, a feeling of powerlessness, a sense of dread or terror, and a feeling of loss of self or of agency. This has been labelled "eco-paralysis"[2]—the inability to act when faced with the complexity and gravity of our situation—which is different from "climate anxiety," just as acrophobia is different from a mere "fear of heights." It is an embodied experience where you lose control and have to look away from the problem in order to focus your senses and then deal with it. The challenge of our time is to find the balance, which involves focusing our attention on the ground while still being aware of the environmental abyss.

From Elements to Interactions

From European Antiquity throughout and past the Middle Ages, 'the elements' referred to air, fire, water, and earth: the four basic substances which everything was assumed to be made of. Some added a more mysterious fifth element: aether, the medium that fills the void between the stars and allows light and gods to travel through a vacuum. Other cultures have had similar concepts. For example, the Tibetan Buddhist medical practice, Sowa-Rigpa, operates with five elements, which all animate and inanimate phenomena are composed of: Earth, water, fire, air, and

[1] Timothy Morton, *Hyperobjects: Philosophy and Ecology after the End of the World* (Minneapolis: University of Minnesota Press, 2013), 3.

[2] Glenn Albrecht, "Chronic Environmental Change: Emerging 'Psychoterratic' Syndromes," in *Climate Change and Human Well-Being: Global Challenges and Opportunities*, ed. Inka Weissbecker (Springer, 2011), 48.

space.[3] These ideas were gradually abandoned in modern science and replaced with other and much longer lists of basic units of the universe such as Antoine Lavoirsier's *Elements of Chemistry*[4] which, among its dozens of chemical substances also lists light and caloric (heat), which we today would describe not as substances but as forms of *energy*—yet another realm of physical categorization.

For centuries, chemical elements were thought to be the irreducible units of matter: atoms that do not consist of simpler substances. That view was challenged in the twentieth century with the discoveries of subatomic units such as protons, neutrons, and electrons—and the list of "elementary particles" that together form what we call "matter" have continued to expand with the inclusion of quarks, photons, bosons, and other strange units, which blur the distinction between mass and energy. Forms of energy include mechanical, chemical, kinetic, and magnetic, but the four fundamental forms are considered to be the gravitational force, the electromagnetic force, and the weak and strong nuclear forces.[5] Energy is not a substance but an interaction, and like with the earlier chemical definitions, these four fundamental interactions are considered to be the ones that cannot be reduced to more basic interactions.[6] These interactions can be described as fields and some of the elementary particles in contemporary physics are simply quanta: units of interaction. This puts an atomic spin to Eugene Gendlin's process model that puts "interaction first,"[7] i.e., before the physical entities. Likewise, drawing from the theory of quantum entanglements, Karen Barad writes that individual entities do not preexist their interactions but "emerge through and as part of their entangled intra-relating."[8]

Interactional fields are thus some of the fundamental elements of the universe. They are what ties things together and are the conditions for things to exist—or, as Gendlin says, "the interactional event determines the individual entities."[9] The strong nuclear force binds quarks together to form the neutrons and protons in the atomic nucleus and the gravitational force, the electromagnetic force regulates the interaction of atoms and molecules that form the objects we observe in the phenomenological space, and the gravitational force keeps the planets and other large

[3] Padma Gurmet, "Sowa-Rigpa: Himalayan Art of Healing," *Indian Journal of Traditional Knowledge* 3, no. 2 (2004).

[4] Antoine Lavoisier, *Elements of Chemistry*: *The Development of Chemistry 1789–1914* (Taylor & Francis, 1790).

[5] And to tie the loop back to antiquity, there have also been theories of a more mysterious fifth force, as described in Allan Franklin and Ephraim Fischbach, *The Rise and Fall of the Fifth Force: Discovery, Pursuit, and Justification in Modern Physics* (Springer, 2016).

[6] Sylvie Braibant, Giorgio Giacomelli, and Maurizio Spurio, *Particles and Fundamental Interactions: An Introduction to Particle Physics* (Springer Science & Business Media, 2011): 109. This passage is also informed by conversations with Thøger Emil Rivera-Thorsen who is an astrophysicist at the University of Stockholm, although I take responsibility for any errors or misconceptions.

[7] Eugene Gendlin, *A Process Model* (Northwestern University Press, 2017), 30.

[8] Karen Barad, *Meeting the Universe Halfway: Quantum Physics and the Entanglement of Matter and Meaning* (Durham: Duke University Press Books, 2007), ix.

[9] Gendlin, *A Process Model*, 38.

objects together and keeps them in their relative orbits around each other while it also makes sure we stay on this planet and know what's down and what's up. It is this latter force, this fundamental interaction we call "gravity," I want to dwell on here. One aspect of the process philosophical view, represented by Gendlin and Barad, in which entities do not exist independent of or prior to their interactions is that we should not view human beings as separate things that experience gravity on a given planet but as entanglements that emerge through a process in which gravity and interaction with this planet is a precondition. We are not just inhabitants *on* planet Earth but beings *of* planet Earth, we are entangled with and emerge from its gravitational field. This also means that when we are disturbing the planetary processes we are not merely "harming Earth" as if it was separate from us; rather, we are changing the processes that are the foundation for our own existence. Thus, living in the Anthropocene requires us to find a new relation to gravity—a new sense of balance.

A Sense of Gravity

In the book *The Sense of Space*, Davis Morris analyzes the importance of the human body being "evolved in a gravitational environment" both for our biological evolution and individual development and for our phenomenological orientation in the world. He gives accounts of astronauts who, in situations of weightlessness in outer space where concepts like up and down are in a sense meaningless, still experience a strong sense of direction which indicates that the Earth is still implied in their body and that, if they attempt to let go of this earthly directedness, they have a phenomenological sense of falling:

> Whatever is beneath the phenomenon, it must be understood in terms of a moving body crossed with earth, a body that carries its relation to Earth along with it. This is what the concept of the topology of residing aims to do: to understand 'up' and 'down' not in terms of the isolate body, but in terms of the body's residential relation to earth.[10]

We are thus, he says, at a neurological and phenomenological level, beings for whom "gravity and thence earth matter."[11] Though this statement is undoubtedly true, I want to add that this relationship—this mattering—between our bodies and gravity/Earth is not always a harmonious one. Our bodies and neurology have evolved to have a sense of space and direction in relation to Earth, but this also means when that orientation to and in the world is broken or disturbed it can have harmful and debilitating effects on our bodies. This is evidenced in people, like me, who suffer from neurological conditions resulting in what we normally, though a bit misleadingly, call fear of heights.

---

[10] David Morris, *The Sense of Space* (Albany: State University of New York Press, 2004), 152.

[11] Morris, 152.

The technical term for the (irrational) fear of heights is *acrophobia* but a related and more specific condition is *basophobia*—the fear of falling. I am not afraid of heights as such, at least not at a cognitive level. I am perfectly fine with flying, and I can be on the top floor of a tall building without breaking a sweat. Until that is, I get close to anything that to my brain looks like an edge that one could theoretically fall off of. It does not matter that I know perfectly well that it is impossible to fall off that edge—perhaps there is just a small crack that wouldn't fit a human body, or maybe there's a railing or a solid but transparent obstacle like armored glass preventing any passage. The knowledge that I am up high is not the problem; it is the visual stimulant caused by the discrepancy between where I am and the world below that causes a reaction in my brain and the rest of my body. Sometimes a great height is not even necessary; positions in the landscape that cause optical illusions of sudden drops can have the same effect. In fact, sitting safely on my couch watching a movie with certain scenes can trigger the body's neurological response almost as if it was actually there.

I remember the first time I had this sensation. I was a child. I don't know how old but maybe around 6 or 7. We were visiting someone who lived on the top floor of a 13-floor building (the tallest in my neighborhood). They had a great balcony which was covered but at the edges of the walls around it, there were small slits in the floor to let rainwater out. As I was playing on that floor, I suddenly noticed the drop through that slit and my body froze, I felt myself needing to get away not just from the slit but also from the wall and I could not come near it again without feeling the dizzying sensation and a nervous suck in my stomach and I did not feel at ease in that apartment again.

In general, though, this condition did not affect me much as a child. It only really manifested itself when I was an adult, starting in my 20ies and accelerating from there. At first, it came as a surprise—like a sudden neurological disorder where I would lose control of my body as I got close to an edge that I just a few seconds before had been walking towards calmly with no fear or anxiety. I did not understand why this happened and it annoyed me immensely as I enjoy hiking and climbing in mountains and elsewhere—an activity that is rendered difficult to impossible with this condition. But as it went on and I became more familiar with the experience I grew to know to expect it, which on the one hand meant that I could prepare for it and take precautions such as approaching slowly and, on the other, meant I would sometimes get anticipatory anxiety by just knowing I was approaching something that my body recognizes as potentially triggering. In either case, it is part of my life and I have learned more or less to deal with it, although it sometimes causes problems—for example when I am excited about a potentially beautiful view or a challenging climb and suddenly get overpowered by the reaction I should have anticipated.

It is difficult to give a phenomenological account of what happens in my body because one of the characteristics is a sort of loss of consciousness and awareness, or rather an inability to focus on certain aspects of the world combined with a hyperawareness of other aspects. When I am suddenly attacked by acrophobia or basophobia my field of attention shrinks: quite literally I feel my field of vision shrinking so I can only see what is right in front of me while the rest of the world gets blurry. There is a sensation of being sucked into a hole and the muscles in my body cramp up in defense—I need to sit down, close my eyes, and shut out the sight that caused the discomfort. I feel my heart rate go up, my breathing gets harder, and I lose control over my body. Only by a conscious effort of relaxing, calm breathing, finding a visual focal point, and deciding on a path and direction, can I gradually regain enough control to move to a location that feels safer—sometimes by using my entire body and staying close to the ground to not lose my balance—although that movement is accompanied with some anxiety, a very narrow visual attention, and partial bodily paralysis.

A fear of falling and a certain apprehension towards steep drops from great heights have probably had some evolutionary advantage, but, needless to say, it is not my experience that this condition helps my survival. On the contrary, it has not only caused annoyance and embarrassment but also made situations more dangerous than they should have been; to suddenly lose your visual awareness and motoric control in situations where those are some of the most important skills hardly promotes my safety and survival—or that of my offspring. Having children has only made it worse, as my attention is naturally focused on them when we are hiking and thus it suddenly becomes divided when an elevational drop enters my field of vision; it is especially this visual division that triggers the sense of vertigo: when the part of my brain that is in charge of knowing where I am in relation to my immediate environment is suddenly interrupted by visual stimuli that challenge its operations and points out that my "immediate" surroundings include a place that is actually much further below. A quick movement of my eyes as one of my children changes location, in a place I otherwise consider safe, might cause such a visual input and induce dizziness and an anticipatory sense of falling.

The point here is that it is the becoming-aware of the abyss, the potential disaster (whether or not it is legitimately dangerous according to a more cognitive consideration), that causes the bodily and mental paralysis. In order to move and act, I need to divert my attention away from the thing that sucks it in, and consciously focus on what I can do: take the next step, and then the next one after that. I need to visualize a trajectory for my next immediate actions. The end-goal is beyond the reach of my awareness, it is a distraction that takes focus away from what must be done here and now. Of course, while I cannot let neither the end-goal nor the abyss take my attention, I still need to keep them in the periphery of my awareness as they are the guides that inform the immediate paths I must focus on. Maintaining this balance of attention can be a tricky feat. These personal reflections resonate with testimonies from others. Geographers Aharoni Avi and Izhak Schnell interviewed several people who, despite suffering from different forms of spatial phobias, give similar accounts.[12] They summarize the phenomenological condition thus:

[12] Aharoni Avi and Izhak Schnell, "Distortion of the Perception of Existential Space: As Illustrated by Phobias," *Psychology and Behavioral Science International Journal* 15, no. 3 (2020): 1–15.

> [T]he space where the experience takes place undergoes a warping of the field or the borders, and the space in general becomes distorted. [...] the edges of the space seem to shrink, compressing the spatial field, while the borders expand until there are no borders at all and the entire spatial field has melted away. [...] a lack of spatial stability leads to feelings of powerlessness in the face of external forces, especially as regards the external space and its borders. The loss of the ability to stabilize our existential space results in feelings of panic. [...] During these moments, the body feels as if it has been separated from all awareness.

Phenomenological Disembodiment

The symptoms of height-induced acrophobia and vertigo are also well-described in the historical literature. There are accounts from ancient Greece, Rome, and China of soldiers becoming dizzy and going weak in the knees when faced with sudden visual drops in altitude,[13] and the eighteenth century British physician Erasmus Darwin devoted several chapters to describing the physiological causes and effects of vertigo which he divided into those that are related to the sense of vision, the sense of hearing, and the sense of touch, with the "vertigo visualis" being caused by a "less perfect vision."[14] Darwin was most likely on the right track, as medical science today tends to agree that "visual height intolerance" can be a result of impaired visual control, or rather an impairment of the vestibular system (which involves both the eyes, the inner ear, and many other parts of the neurological and muscular system) that makes visual input play a greater role in the maintenance of balance.[15] Whatever causes the condition, it seems to be a complex somatic disorder that is both historically normal and prevalent today: it is the most common phobia, taking up 41% of all self-reported phobias in Iceland, and as it does have an evolutionary function in helping especially children to avoid dangerous situations,[16] it is not surprising that it has both environmental and genetic components.[17]

---

[13] Thomas Brandt et al., "Susceptibility to Fear of Heights in Bilateral Vestibulopathy and Other Disorders of Vertigo and Balance," *Frontiers in Neurology* 9 (2018): 406; Thomas Brandt and Doreen Huppert, "Fear of Heights and Visual Height Intolerance," *Current Opinion in Neurology* 27, no. 1 (2014): 111–17; Matthias Bauer, Doreen Huppert, and Thomas Brandt, "Fear of Heights in Ancient China," *Journal of Neurology* 259, no. 10 (2012): 2223–25; Doreen Huppert et al., "Fear of Heights in Roman Antiquity and Mythology," *Journal of Neurology* 260, no. 9 (2013): 2430–32.

[14] Erasmus Darwin, *Zoonomia, or, The Laws of Organic Life* (Philadelphia: Printed by T. Dobson, 1797), 309, 119.

[15] Doreen Huppert, Max Wuehr, and Thomas Brandt, "Acrophobia and Visual Height Intolerance: Advances in Epidemiology and Mechanisms," *Journal of Neurology* 267, no. Suppl 1 (December 2020): 231–40; Mark S. Redfern, Lucy Yardley, and Adolfo M. Bronstein, "Visual Influences on Balance," *Journal of Anxiety Disorders*, The Interface of Balance Disorders and Anxiety, 15, no. 1 (2001): 81–94.

[16] Huppert, Wuehr, and Brandt, "Acrophobia and Visual Height Intolerance."

[17] Zuzanna Misiewicz et al., "A Genome-Wide Screen for Acrophobia Susceptibility Loci in a Finnish Isolate," *Scientific Reports* 6 (2016): 39345.

This raises the question of whether it should be seen as a phobia in the first place, rather than a result of natural neurological and physiological diversity. As one medical paper writes: “Though height vertigo is generally regarded as a psychopathological process, we hypothesize that it might instead result from an intersensory mismatch when visual information is at variance with vestibular and proprioceptive inputs”; it emphasizes that “physiological 'distance vertigo' must be distinguished from psychological 'acrophobia'.”[18] In other words, I am mentally fine—it is just a problem of communication between my visual neurons, inner ear, and motoric system, not a problem with my psyche! This might also explain why, while it is clear from a survey of literature that it is widely studied by the medical sciences, it has received less philosophical attention than some of the other place and environment-related phobias, such as agoraphobia which has a body of philosophical literature including Dylan Trigg’s *Topophobia: A Phenomenology of Anxiety*, Paul Carter’s *Repressed Spaces: The Poetics of Agoraphobia*, and Joyce Davidson’s *Phobic Geographies: The Phenomenology and Spatiality of Identity*.[19] I don’t think we should let my psyche off the hook that easily. Acrophobia has a psychological component and it deserves philosophical investigation. First of all, the neurological and sensory condition cannot be separated from the psychological phobia. As I have described, the experience with my bodily reactions in certain environmental situations results in an anticipatory psychological attitude: I might not cognitively fear the coming height, but I do fear my body’s reaction to it, so I want to own the word *phobia* when describing my condition. Secondly, the somatic experience of sensory perception and how it shapes our component in the world is itself a topic of philosophical investigation. And thirdly, the condition does share a lot of similarities with those described by sufferers of agoraphobia, and I will therefore draw upon some of these texts.

One philosophical problem that must be addressed already is the language I have been using in phrases like “I fear my body’s reaction.” This problematic language implies a separation between my self and my body, which in one sense we should reject but which for the phobic can be a real phenomenological experience. I have described that I might feel the bodily sensation of danger and vertigo even in situations where I cognitively know that there is no danger. The same is true for the agoraphobic who, as Trigg writes, can be gripped by a sense of danger in a given place “despite the fact that I objectively know little harm could come to me in such an environment” which leads to the feeling that “My body appears to be speaking on behalf of a threat that ‘I’ cannot detect.”[20] Trigg says that “At the heart of the

[18] Willem Bles et al., “The Mechanism of Physiological Height Vertigo: II. Posturography,” *Acta Oto-Laryngologica* 89, no. 3–6 (1980): 534–40.

[19] Paul Carter, *Repressed Spaces: The Poetics of Agoraphobia* (London: Reaktion Books, 2004); *Dylan Trigg, Topophobia: A Phenomenology of Anxiety* (Bloomsbury Publishing, 2016); Joyce Davidson, *Phobic Geographies: The Phenomenology and Spatiality of Identity* (Routledge, 2017).

[20] Dylan Trigg, “Bodily Moods and Unhomely Environments: The Hermeneutics of Agoraphobia and the Spirit of Place,” in *Interpreting Nature: The Emerging Field of Environmental Hermeneutics*, ed. Forrest Clingerman et al. (Fordham University Press, 2013), 171.

phobic interpreting of the environment is thus a dissonance between cognitive and mental phenomena."[21] The fact that I know there is no danger, and yet my body reacts as if there, results in frustration and a sense of disconnection between me as "the rational being" and me as the irrational bundle of neurons that is "my body." If I can lose control over my body, which is what it feels like, then what is this "I" that should be in control? According to Trigg, this "supposed irrationality only gains a currency if we grant rational thinking an ontological primacy over the body's experience of the world."[22]

Uncomfortable Space

The first part of the word *acrophobia* comes from the Greek *ἄκρον* (akron) which means summit or peak, as in the top of a mountain. It is thus quite literally an experience related to certain spaces and environments. There are many other types of spatial and environmental phobias, such as the fear of open spaces or agoraphobia which is analyzed extensively by the philosopher Dylan Trigg who also draws upon his own experience with the condition.[23] Others include claustrophobia, the fear of enclosed spaces, hylophobia, the fear of forests, and thalassophobia, the fear of the ocean, as well as more specific ones such as the fear of crossing bridges (gephyrophobia) or streets (agyrophobia). Trigg uses the name "topophobia" (fear of places) to refer to this broad category of spatial phobias or anxiety-inducing relations a person might have with specific places.[24] There are two implications of this concept: (1) there is an environmental component as these conditions affect and disrupt our experience—and appreciation—of nature or other spaces, and (2) the relation to space is also a relation to our bodies: my discomfort with abysses is primarily a discomfort with my own body in the situation.

Regarding the latter implication, we must draw upon Maurice Merleau-Ponty who says that "our body is not merely one expressive space among the rest" but rather "our general medium for having a world."[25] It is through our bodily interaction with and moving around in the world that we get a sense of the world and by doing this we also get a sense of our bodies as things that can move in the world and change perspectives. Thus, he says both "I am conscious of my body via the world" and "I am conscious of the world through the medium of my body."[26] For some, this reciprocal relation between body and world is seen as a means for appreciating nature, such as when Guðbjörg R. Jóhannesdóttir and Sigridur Thorgeirsdottir urge

[21] Trigg, 172.

[22] Trigg, 172.

[23] Dylan Trigg, *Topophobia: A Phenomenology of Anxiety* (Bloomsbury Publishing, 2016).

[24] Trigg, *Topophobia*, xxi.

[25] Maurice Merleau-Ponty, *Phenomenology of Perception* (Routledge, 2005), 169.

[26] Merleau-Ponty, 94–95.

us to "reclaim nature" by "reclaiming the body,"[27] while for others, like Herbert Schroeder, it means that being in nature can be used for adjusting our intrinsic bodily processes and achieve relaxation and serenity.[28] This seems very harmonious, but I want to explore what happens when there is a misalignment between the place and our bodies: when the space we are in seems hostile and this sense is inseparable from the sense that there is a conflict going on within our own bodies. By losing our sense of space we also loose the sense of self; we become split in a powerless Cartesian self with no extension and little influence on the world of extended space, *res extensa*, and a material body, *res extensa*, that is in and of the space but feels alien and beyond our control. If we give in to that feeling completely, then nature is not something to be appreciated, respected, and protected but rather something we must guard ourselves against by building a safe space where our selves can achieve coherence and stability. This feeling, this desire for control and for self-control, is what the institutions of the state and so-called "civilization" has always been for, and in its extreme, it devolves into a generalized ecophobia.[29]

Merleau-Ponty writes that "to be a body, is to be tied to a certain world [...] our body is not primarily *in* space: it is *of* it."[30] But we are not all of the same place and different places affect us differently. This can result in different bodily feelings in different spaces. These might not exactly be phobic experiences but something more subtle: a sense of discomfort in your body and the sense that a place is wrong in ways that are hard to articulate. For example, as an immigrant to Iceland, when I'm driving around the countryside and see the summer cabins along the road placed out in nowhere on barren, rocky fields with nothing to shelter them, I have an eerie sense of wrongness and anxiety. I feel there is no protection, nothing that can make these places homely—like the sky could just swallow you. I know many others feel this too, and from having taught international students I know many of us feel a deep discomfort by the general absence of trees. We don't notice this until we find a forest and suddenly feel the sense of calm and home that our bodies have been missing. Many Icelanders, on the other hand, have an instinctive dislike for forests and tall trees. In conversations, I have tried to circle in on what that might be about and a thing that frequently comes up is a discomfort about their vision being obstructed and a disorientation from the horizon being broken by vertical objects. This is not unique to Icelanders. Danes who grew up along the vast coastline of Western Europe also express a feeling akin to claustrophobia when they do not have access to the vast horizon. Similarly, others I have spoken to have described discomfort and disorientation when moving to a place without mountains to use for navigation and being grounded in place. Our bodies are never just *in* a space but *of* a space,

---

[27] Guðbjörg R. Jóhannesdóttir and Sigridur Thorgeirsdottir, "Reclaiming Nature by Reclaiming the Body," *Balkan Journal of Philosophy* 8, no. 1 (2016): 39–48.

[28] Herbert W. Schroeder, "The Felt Sense of Natural Environments," *The Folio* 21, no. 1 (2008): 63–72.

[29] Steven Hartman and Patrick Degeorges, "'DON'T PANIC': Fear and Acceptance in the Anthropocene," *ISLE: Interdisciplinary Studies in Literature and Environment*, 2019.

[30] Merleau-Ponty, *Phenomenology of Perception*, 171 (italics in original).

which is why migrating to and settling into a new and different space can be a difficult process.

Drawing upon Martin Heidegger's concept *Stimmung* (mood), Trigg asks the question of how our bodily moods affect our experience of environments: "To do justice to the nature of environmental experience, we need to address the relationship between the body's moods and the variations in environmental interpretation that arise from those moods."[31] Our past experiences are stored in our bodies and influence how we feel in an environment but the mood can also be shaped by an expectation about the future, which is shaped by the past, such as when the acrophobe's experience of a landscape is influenced by the expectation that he will soon come around a corner that might reveal an abyss that triggers the condition—the condition is already active in that situation as an apprehensive nervousness that affects the mood of the experience, making the whole landscape feel a bit hostile and threatening.

The environmental philosopher Angelika Krebs also uses the Heideggerian *Stimmung* in her exploration of why landscape beauty matters. She argues that the experience of beautiful landscapes makes us feel at home in the world, "not only because they have a pleasant physiological and psychological impact on us, but also because they indicate, by the way they look, sound, and smell, that they can support humans and provide for their needs."[32] This argument is problematic when applied either to people whose problem is exactly this of not feeling at home, i.e., not feeling safe and comfortable, in certain landscapes no matter how beautiful they might be, or to landscapes which—despite being beautiful—do not really provide anyone with this sense that they can provide for human needs. Annually, millions of tourists come to Iceland to experience the landscapes here: magnificent waterfalls, rolling hills, dramatic cliffs, rocky fields of petrified lava, smoking volcanoes, stony highland deserts, and retreating glaciers. It is indeed beautiful, but it is one thing to see them as a tourist and another to consider them your home because while they might fill you with awe, one thing they do not do is indicate that they are here to support humans and provide for their needs. Some landscapes simply are uncanny. They feel alienating and uninviting—even to the local population. As if nature is telling you that this is not a place where humans can comfortably live; life here will be a struggle. And maybe that has a value too as a reminder that the world does not exist to please our needs, but is still, as the Icelandic philosopher Páll Skúlason says, an objective reality and the foundation of our existence.[33]

[31] Trigg, "Bodily Moods and Unhomely Environments," 163.

[32] Angelika Krebs, "Why Landscape Beauty Matters," *Land* 3, no. 4 (December 2014): 1263.

[33] "Mediatations at the Edge of Askja" in Páll Skúlason, *Reflections on Nature* (Háskólaútgáfan, 2019).

Staring into the Abyss, or, Eyes on the Horizon

My neurological condition makes me unable to stare into the abyss; I lose my sight and my epistemological world shrinks, my body recoils and diverts my eyes. Is this not similar to what many of us are feeling and doing regarding the environmental crises? I don't just mean the climate deniers who deliberately look away to suppress the sense of anxiety but all of us: None of us can fully see or comprehend the reality of these crises. Not even climate scientists know what will happen to the state of the planet when we cross the threshold of 3-degree warming towards which we are currently headed. The possibility of a planet that is inhospitable to human life is surely enough to make us divert our gaze lest we go weak in the knees and lose control.[34] In a global survey of climate-related emotions among young people, 75% said that the future is frightening, and a significant number feel they have no future.[35] I think part of this is related to the epistemological fact that we don't know what the future will look like—we are in a situation where we cannot imagine a future with ourselves in it. This makes it hard to navigate in the present. We are at a cliff where the dizzying abyss makes us paralyzed. Glenn Albrecht, who researches psycho- and somaterratic syndromes, says: "As uncertainty rises, I suggest that anxiety might turn into a form of helplessness or paralysis with the inability to make definitive decisions about what to do next entrenched in a non-predictable world."[36] Inaction might therefore not, as Renée Lertzman points out, be a result of caring too little, but of "caring too much."[37] We don't know what to do. We feel powerless. When we stare into the abyss, we either become paralyzed with fear or we divert our eyes, our world shrinks and we either desperately focus on something we can do, no matter how inefficient (like using paper straws), or on living in the moment while constantly keeping the lurking dread at bay.

Sigmund Freud wrote that in phobias "we often find the recollection of an anxiety attack; and what the patient actually fears is the occurrence of such an attack under the special conditions in which he believes he cannot escape it."[38] This is certainly the case for my acrophobia, which is not merely the neurological condition that is triggered with the right visual and environmental stimuli but also my bodily recollection of past experiences of the condition and the awareness that it might occur again when I am in a similar environment. It is thus, he says, "the

[34] Will Steffen et al., "Trajectories of the Earth System in the Anthropocene," *Proceedings of the National Academy of Sciences* 115, no. 33 (2018).

[35] Caroline Hickman et al., "Climate Anxiety in Children and Young People and Their Beliefs about Government Responses to Climate Change: A Global Survey," *The Lancet Planetary Health* 5, no. 12 (2021).

[36] Albrecht, "Chronic Environmental Change."

[37] Renée Lertzman, "The Myth of Apathy," *The Ecologist* 38, no. 5 (2008).

[38] Sigmund Freud, *The Standard Edition of the Complete Psychological Works of Sigmund Freud*, trans. James Strachey, vol. III (London: The Hogarth Press, 1962), 81.

preparedness for anxiety, which constitutes the nucleus of the neurosis."[39] In other words, the defining feature of the phobia is not the somatic state in the situation that is feared but the anticipatory fear of that somatic state: The mind is preparing itself for the anticipated traumatic event and that preparation can sometimes be less than useful if it makes the body unable to act in the present. Is this not what many of us are going through each summer as we digest the news about hurricanes, floods, droughts, and wildfires? We are not merely terrified of these events themselves but see them as harbingers of an even more terrifying—and yet, unknown—future event. Drawing on Freud, Timothy Morton says that we might have a certain pleasure in absorbing information about global warming because the anticipatory fear is less intense than the fear you experience when you are in the middle of the traumatic event. Thereby we are "trying to install ourselves at a fictional point in time *before global warming happened*."[40] In a way, we are suffering from "pre-traumatic stress disorder"[41] but thereby also avoiding the fact that we are in fact in the middle of the event.

Elsewhere I wrote that "a fixation upon a particular vision of the future, as if it is predetermined by the past, can constrain our conceivable options in the present and thus make that future inevitable."[42] To take a mundane example: when my child was learning to ride a bike, I noticed that he would focus his gaze upon the obstacles he needed to avoid, but as the eyes and the motoric system are coordinated it meant he would steer himself directly towards those obstacles. To bike, you need to somehow keep your focus on where you want to go while also being aware of what you want to avoid by keeping it in your peripheral vision. This is a balancing act that requires practice. This also applies to my fear of heights: If I look into the abyss, my body and sight recoil and I lose my balance. To overcome that paralysis, I need to focus on my immediate surroundings as my world shrinks and then gradually take my eyes to the horizon—i.e., the place where I want to go. And yet, it is still necessary to keep the abyss in the peripheral vision in order to know what to avoid. Faced with planetary crisis and a feeling of impending doom we all need to practice this balance of focus.

Balance is one of the human senses which we could only have developed because we are bodily beings that evolved on a planet with a specific gravity. If we want to survive on Earth we need to practice our balancing skills—both our individual balance so as to neither be overcome by paralyzing fear nor foolish and cruel optimism,[43] and the socio-economic balance required for us to make a life on this planet without destroying the foundations for that life by overreaching the ecological and planetary

[39] Freud, III:125.

[40] Timothy Morton, *Being Ecological* (Cambridge, Massachusetts: MIT Press, 2019): xxiii (italics in original).

[41] E. Ann Kaplan, "Is Climate-Related Pre-Traumatic Stress Syndrome a Real Condition?," *American Imago* 77, no. 1 (2020).

[42] Ole Martin Sandberg, "Climate Disruption, Political Stability, and Collective Imagination," *Radical Philosophy Review* 23, no. 2 (2020): 331–60.

[43] Lauren Berlant, *Cruel Optimism* (Duke University Press, 2011).

tipping points.[44] As the scientists who are working to quantify the planetary boundaries make clear, these boundaries should not be seen as separate processes that can be analyzed or dealt with in isolation from the others. They are, rather, interacting parts of the Earth process and going beyond the boundary of one will also impact the others, potentially triggering a planetary spiraling out of control in which "the Earth system cannot continue to function in a stable, Holocene-like state."[45] It is not just our sense of balance that is disturbed, but also the balance of the planet that is the source of our sense of balance. We must face the gravity of our situation and re-become Earthly beings.

Bibliography

Albrecht, Glenn. 2011. Chronic Environmental Change: Emerging 'Psychoterratic' Syndromes. In *Climate Change and Human Well-Being: Global Challenges and Opportunities*, ed. Inka Weissbecker, 43–56. Springer.

Avi, Aharoni, and Izhak Schnell. 2020. Distortion of the Perception of Existential Space: As Illustrated by Phobias. *Psychology and Behavioral Science International Journal* 15 (3): 1–15.

Barad, Karen. 2007. *Meeting the Universe Halfway: Quantum Physics and the Entanglement of Matter and Meaning*. Durham: Duke University Press Books.

Bauer, Matthias, Doreen Huppert, and Thomas Brandt. 2012. Fear of Heights in Ancient China. *Journal of Neurology* 259 (10): 2223–2225.

Berlant, Lauren. 2011. *Cruel Optimism*. Duke University Press.

Bles, Willem, Theo S. Kapteyn, Thomas Brandt, and Friedrich Arnold. 1980. The Mechanism of Physiological Height Vertigo: II. Posturography. *Acta Oto-Laryngologica* 89 (3–6): 534–540.

Braibant, Sylvie, Giorgio Giacomelli, and Maurizio Spurio. 2011. *Particles and Fundamental Interactions: An Introduction to Particle Physics*. Springer Science & Business Media.

Brandt, Thomas, and Doreen Huppert. 2014. Fear of Heights and Visual Height Intolerance. *Current Opinion in Neurology* 27 (1): 111–117.

Brandt, Thomas, Eva Grill, Michael Strupp, and Doreen Huppert. 2018. Susceptibility to Fear of Heights in Bilateral Vestibulopathy and Other Disorders of Vertigo and Balance. *Frontiers in Neurology* 9: 406.

Carter, Paul. 2004. *Repressed Spaces: The Poetics of Agoraphobia*. London: Reaktion Books.

Darwin, Erasmus. 1797. *Zoonomia, or, The Laws of Organic Life*. Philadelphia: Printed by T. Dobson.

Davidson, Joyce. 2017. *Phobic Geographies: The Phenomenology and Spatiality of Identity*. Routledge.

Franklin, Allan, and Ephraim Fischbach. 2016. *The Rise and Fall of the Fifth Force: Discovery, Pursuit, and Justification in Modern Physics*. Springer.

Freud, Sigmund. 1962. *The Standard Edition of the Complete Psychological Works of Sigmund Freud*, Trans. J. Strachey. Vol. III. London: The Hogarth Press.

Gendlin, Eugene. 2017. *A Process Model*. Northwestern University Press.

Gurmet, Padma. 2004. Sowa-Rigpa: Himalayan Art of Healing. *Indian Journal of Traditional Knowledge* 3 (2).

[44] UNDRR, "Global Catastrophic Risk and Planetary Boundaries: The Relationship to Global Targets and Disaster Risk Reduction" (United Nations Office for Disaster Risk Reduction, 2022), 7.

[45] Johan Rockström et al., "A Safe Operating Space for Humanity," *Nature* 461, no. 7263 (September 2009): 474, https://doi.org/10.1038/461472a

Hartman, Steven, and Patrick Degeorges. 2019. *'DON'T PANIC': Fear and Acceptance in the Anthropocene*. ISLE: Interdisciplinary Studies in Literature and Environment.

Hickman, Caroline, Elizabeth Marks, Panu Pihkala, R. Susan Clayton, Eric Lewandowski, Elouise E. Mayall, Britt Wray, Catriona Mellor, and Lise van Susteren. 2021. Climate Anxiety in Children and Young People and Their Beliefs about Government Responses to Climate Change: A Global Survey. *The Lancet Planetary Health* 5 (12).

Huppert, Doreen, Judy Benson, Benedikt Krammling, and Thomas Brandt. 2013. Fear of Heights in Roman Antiquity and Mythology. *Journal of Neurology* 260 (9): 2430–2432.

Huppert, Doreen, Max Wuehr, and Thomas Brandt. 2020. Acrophobia and Visual Height Intolerance: Advances in Epidemiology and Mechanisms. *Journal of Neurology* 267 (Suppl 1): 231–240.

Jóhannesdóttir, Guðbjörg R., and Sigridur Thorgeirsdottir. 2016. Reclaiming Nature by Reclaiming the Body. *Balkan Journal of Philosophy* 8 (1): 39–48.

Kaplan, E. Ann. 2020. Is Climate-Related Pre-Traumatic Stress Syndrome a Real Condition? *American Imago* 77 (1).

Krebs, Angelika. 2014. Why Landscape Beauty Matters. *Land* 3 (4): 1251–1269.

Lavoisier, Antoine. 1790. *Elements of Chemistry. The Development of Chemistry 1789–1914*. Taylor & Francis.

Lertzman, Renée. 2008. The Myth of Apathy. *The Ecologist* 38 (5).

Merleau-Ponty, Maurice. 2005. *Phenomenology of Perception*. Routledge.

Misiewicz, Zuzanna, Tero Hiekkalinna, Tiina Paunio, Teppo Varilo, Joseph D. Terwilliger, Timo Partonen, and Iiris Hovatta. 2016. A Genome-Wide Screen for Acrophobia Susceptibility Loci in a Finnish Isolate. *Scientific Reports* 6: 39345.

Morris, David. 2004. *The Sense of Space*. Albany: State University of New York Press.

Morton, Timothy. 2013. *Hyperobjects: Philosophy and Ecology after the End of the World*. Minneapolis: Univ Of Minnesota Press.

———. 2019. *Being Ecological*. Cambridge, MA: MIT Press.

Redfern, Mark S., Lucy Yardley, and Adolfo M. Bronstein. 2001. Visual Influences on Balance. *Journal of Anxiety Disorders, The Interface of Balance Disorders and Anxiety* 15 (1): 81–94.

Rockström, Johan, Will Steffen, Kevin Noone, F. Åsa Persson, Stuart Chapin, Eric F. Lambin, Timothy M. Lenton, et al. 2009. A Safe Operating Space for Humanity. *Nature* 461 (7263): 472–475. https://doi.org/10.1038/461472a.

Sandberg, Ole Martin. 2020. Climate Disruption, Political Stability, and Collective Imagination. *Radical Philosophy Review* 23 (2): 331–360.

Schroeder, Herbert W. 2008. The Felt Sense of Natural Environments. *The Folio* 21 (1): 63–72.

Skúlason, Páll. 2019. Reflections on Nature. *Háskólaútgáfan*.

Steffen, Will, Johan Rockström, Katherine Richardson, Timothy M. Lenton, Carl Folke, Diana Liverman, Colin P. Summerhayes, et al. 2018. Trajectories of the Earth System in the Anthropocene. *Proceedings of the National Academy of Sciences* 115 (33).

Trigg, Dylan. 2013. Bodily Moods and Unhomely Environments: The Hermeneutics of Agoraphobia and the Spirit of Place. In *Interpreting Nature: The Emerging Field of Environmental Hermeneutics*, ed. Forrest Clingerman, Brian Treanor, Martin Drenthen, and David Utsler. Fordham University Press.

———. 2016. *Topophobia: A Phenomenology of Anxiety*. Bloomsbury Publishing.

UNDRR. 2022. *Global Catastrophic Risk and Planetary Boundaries: The Relationship to Global Targets and Disaster Risk Reduction*. United Nations Office for Disaster Risk Reduction.

Ole Martin Sandberg is a Doctor of Philosophy from the University of Iceland where he teaches environmental ethics and philosophy of nature. He has participated in research on Embodied Critical Thinking, microphenomenology, and nature experiences. He also works as a researcher at the Icelandic Museum of Natural History where he works on the concept of biodiversity and intra-species diversity from a process philosophical perspective. He has contributed chapters on Peter

Kropotkin and Richard Dawkins for the volume *Unchaining Solidarity: On Mutual Aid and Anarchism with Catherine Malabou* (Rowman & Littlefield 2021), and on making sense of failures to communicate in contemporary political discussions for the volume *The Psychology and Philosophy of Eugene Gendlin: Making Sense of Contemporary Experience* (Routledge 2023), and published papers on the politics and effects of climate changes.

Chapter 9
Vegetal Pedagogy

Michael Marder

Abstract This chapter considers the possibility of a radical vegetal pedagogy, understood in terms of a learning from plants. After outlining the fundamental sense of this pedagogy, I show (1) how it derives from plant-thinking, and (2) what the possibilities of plant listening hold in store for it. I conclude with a proposal to realign human thinking, perception, and modes of being in the world in keeping with vegetal existence.

Keywords Plant-thinking · Intelligence · Pedagogy · Perception · Listening · Interpretation

Earth-Children

In a beautiful diary note titled "The Earth and Children" and dated July 1876, Fyodor Dostoyevsky writes: "I do not draw any differentiation between the earth and children."[1] Like trees and grasses, bushes and flowers, children are the produce of the earth in an original take on the Biblical creation narrative, according to which Adam and Eve are made from the earth itself. Pedagogy becomes a matter of

[1] Fyodor Dostoyevskiy, "The Earth and Children." *Diaries, July–August 1876*. https://fedordostoevsky.ru/works/diary/1876/07-08/13/. Here and throughout this text all translations from the original Russian are mine.

M. Marder (✉)
Department of Philosophy, The University of the Basque Country (UPV-EHU), Vitoria-Gasteiz, Spain

Ikerbasque, Basque Foundation for Science, Bilbao, Spain

L. Škof et al. (eds.), *Elemental-Embodied Thinking for a New Era*, Sophia Studies in Cross-cultural Philosophy of Traditions and Cultures 42,
https://doi.org/10.1007/978-3-031-42119-8_9

cultivation, of unobtrusively channeling children's growth—physical and psychic—in a way similar to the cultivation of orchard trees or fields of cereals.

In the grand narrative of human history, as Dostoyevsky envisions it, the last word of every civilization, the final "humanization of everything human,"[2] is the unconditional openness to the natural world. Beyond even the confines of cultivation and what Dostoyevsky derides as good manners, is the urge "to expand, to dissolve in, to open toward nature completely, to move toward this golden solar ray, which shines upon us, sinners, from the blue sky, without discerning whether we are worth it or not."[3] (Freud called this urge "the oceanic feeling.") And it is because the last word of civilization is this openness without reserve, the trajectory of human history passes from castles through cities to orchards, as Dostoyevsky has it. So, in 1876, the Russian novelist writes: "There will be an Orchard. Mark my word in one hundred years and recall what I have told you in an artificial garden of this summer resort, among artificial people. Humanity will be renewed in an Orchard and it will be straightened up by an Orchard—that is the formula."[4]

Without a doubt, the reference to the straightening up of humanity is an allusion to Immanuel Kant's famous dictum, preceding Dostoyevsky by nearly a century, that "from the crooked timber of humanity, nothing entirely straight can be made." Yes, nothing straight will come out of crooked timber, but what if the matter, the materiality, the matrix of humanity is not dead wood (timber) but the living woods or orchards that never correspond to the dubious ideal of straightness? Surely, an apple tree that, as a sapling, begins growing crookedly can be straightened up with the help of various supports that from all sides guide and channel its future development. Except that Dostoyevsky does not really speak of the growth of an individual human or tree but about humanity's renewal and straightening by an Orchard, by the culture-nature or nature-culture that it will have become. What does this mean?

One crucial measure of human maturity is the way we treat our environment. Nearly 150 years after the prophetic diary entry, we are closer to transforming the world into a gigantic dump than into a flourishing Orchard. A careless and destructive approach toward the world, which is usually conceived as a kind of playground for the enactment of our phantasies, is irresponsible and childish in the worst sense, forgetful of the strange non-differentiation between children and the earth we've just mentioned. This approach shows no respect for other forms of life, a lack of concern with the future, and the inability to think and to grow beyond the demands of sheer physical survival, itself detrimental if it proceeds at the expense of the milieu, in which one is meant to survive.

Historically, there has been little change in the direction of a more adult behavior toward the environment, despite all talk about "green energy transition" and other such greenwashing measures. Among other living beings, plants have been

[2] Fyodor Dostoyevskiy, "Civilization's Very Last Word." *Diaries, July–August 1876.* https://fedordostoevsky.ru/works/diary/1876/07-08/04/

[3] Fyodor Dostoyevskiy, "What Helps at Thermal Baths? Water or Good Manners?" *Diaries, July–August 1876.* https://fedordostoevsky.ru/works/diary/1876/07-08/10/

[4] Dostoyevskiy, "The Earth and Children." https://fedordostoevsky.ru/works/diary/1876/07-08/13/

particularly mistreated as a result of this attitude because they have been thought of as infinitely malleable matter, on which human form could be stamped or imposed, generally to the detriment of their own biological life. Indeed, Aristotle, who was the first to come up with the notion matter in the West, derived it from the common Greek word for "wood." Like plants, matter was supposed to be a passive receptacle for the form that was, in many cases, alien to vegetal life, not cultivated along with the matter itself. Although Aristotle was still attentive to living forms, after him, a tree converted into a table or a bed became the preferred example of formed matter, while the self-formation of the tree, amenable to patient cultivation and care, was dismissed.

When it comes to respect toward the environment, we are still children, or even infants. More than that, we are terrible, unruly children because, for the most part, we are not open to being educated on the subject. Only punishments, in the shape of natural disasters attributable to global warming or pandemics, have had some (albeit very limited) effect on human behavior, awakening in us a consciousness of the negative consequences that accompany immature environmental conduct. Still, a genuine change of attitudes is unlikely as a result of threats and punishments alone. What is sorely needed is an environmental pedagogy—not one formulated by our fellow humans, but one imparted the world we inhabit.

Pedagogy literally means "the guidance or leadership of children." If we are all children in relation to the environment (both as dependent creatures who are much younger than it and as the immature agents of its destruction), then we ought to be led or guided by the more developed wisdom of the other-than-human world. And, since plants are some of the oldest forms of life on earth, indispensable to the survival of all other living beings, their guidance is especially conducive to an environmental pedagogy.

In a way, we should follow the plants, not only in the evolutionary sense but also in the pedagogic one, seeking guidance from them. There is plenty of resistance to such an injunction, which may sound absurd. More often than not, we pay no attention to the meanings of the vegetal world. We do not listen to what it says to us. And yet, its teaching can be propitious to our own life and our manner of behaving toward the others and toward the world in general. The crooked sapling that is the human stands a chance to grow better, to thrive and to flourish, only if it is propped up, supported on all sides not by wooden or concrete beams, but by trees and shrubs and even weeds.

Plant-Perception

My wager over the past years has been that plant-thinking could become one of the living guides in a more-than-human pedagogy. Although plant intelligence does not amount to plant-thinking, it, too, is a contested term within twenty-first century scientific circles. The conservative tendency is to acknowledge a wide array of plant capacities to detect minute variations in their environments and to react accordingly—as no more than a hodgepodge of biosensors. The distinction between intelligence and thinking runs in parallel to the difference between sensing and perceiving. A thermometer senses temperature variations but does not perceive them. To perceive something is not only to interpret the information the senses supply but also *to be* and *to live* that interpretation, continually incorporating it into the perceiving body itself, to the point where it cannot be removed from that body without its removal inflicting irreparable harm. A thermostat, in its turn, may be intelligent if it uses the information from its thermal sensors for the sake of fulfilling a certain function, such as keeping temperatures in the house steady. But it will not think because it does not exist "for nothing" at the edges of its manufactured body, does not articulate the extremes, does not *dwell* in the intermediate space between itself and its milieu.

It follows from this brief comparison that sensing and intelligence belong together in the paradigm of information processing that continues to rely on the mechanics of the mind-body split. The sensors and the information they detect are not internally interrelated; the link between them is merely external and indifferent to the participating terms. Perception and thinking, conversely, are embroiled with one another through the living body that forges relations to itself and to its world. They are the interfaces between a living body and its environment. More accurately, they are the same interface viewed under different aspects or different edges of the same surface. With regard to plants, the crucial capacities, therefore, are those of perception and thinking, not of sensation and intelligence. However, virtually all the literature on the topic has been concerned, albeit with a fair share of controversy, with the second couple of phenomena. In and of itself, this fact speaks volumes about the state of the art in approaches to plant subjectivity, and, to the subjectivity of other forms of life and non-life, if one throws artificial intelligence into the mix.

In the conclusion to his influential book *What a Plant Knows?* plant scientist Daniel Chamovitz veers on the side of caution. Happy to talk about plant awareness and a glut of things plants know, on matters of intelligence Chamovitz remains unconvinced. According to him, since plants have no brain, cannot suffer, and are unable to engage in a two-way "flow of meaning" with us, every attempt to attribute intelligent behavior to them is doomed to anthropomorphic hyperbole.[5] If plants are not intelligent, then they are even less capable of thinking. But the status of knowledge and awareness that Chamovitz is willing to attribute to plants is woefully hazy. What are their presumably unintelligent knowledge and unthinking awareness? Between the lines of the book's concluding chapter, the former stands for information, while the latter is sensation. Aggravated by a conflation of anatomical structures (the brain) with the functions they perform (cognition), the coding of plant "knowledges" in these terms ensures the survival of metaphysical stereotypes insisting on the deficiency of the vegetal. Qualitatively assessed, the answer to the question Chamovitz poses in the title is not at all new; it is merely embellished with *faits*

[5] Daniel Chamovitz, *What a Plant Knows? A Field Guide to the Senses of Your Garden and Beyond* (Oxford: Oneworld, 2012), 170–3.

divers drawn from the latest research, occasionally making the news, on how plants register or process information.

Another strategy to adopt could be one that, instead of aiming "high"—at intelligence, knowledge, or thinking—we would more modestly take its bearings from perception. After all, perception is the middle term between sensation and thinking as conventionally understood. And, from its place in the middle, it is favorably disposed to the comprehension of plant subjectivity.

Perception consists in the perceptual apparatus and the process it mediates. It includes the perceiving sense organs and what they perceive, with the ongoing interpretation of the world built into the relation. But this analytic description is wanting. Perception is reduced to a nearly mechanical operation unless we concentrate on the fact of sentient embodiment it fleshes out. "The perceptual apparatus" is the living body itself, exposed to the outside and procuring meaning from this exposure. The difference between hardware and software that, in the computational paradigm, resurrects the infamous split does not apply. To remove the complex of perception from a body is to convert it into another kind of body, a corpse, even if its basic vital functions (e.g., breathing) are ensured, as in a persistent vegetative state.

Plant perception is a case in point of how the entire body is involved in perceptual interactions. Plant organs are sense organs, each of them responsible for several senses at once. As a consequence, plants probably perceive the world in a synesthesia, where various kinds of the perceived (touched, seen, heard…) merge, even as the types of perceiving (touching, seeing, hearing…) converge on the same organ. More than that (and this is where plant-thinking springs up from plant-perceiving), the interpretive nexus embedded in the manifold perceptual relation is an intricately tied knot of several hermeneutical acts. Rather than translate it into multivariable algorithmic computation, as my colleague Francisco Calvo Garzón does,[6] it is necessary to accompany its dynamics phenomenologically, deducing how plant body itself thinks by relating to its world and to itself. Lived vegetal hermeneutics must not lose sight of the fact of embodiment—how plant bodies undergo the processes of growth, decay, metamorphosis—which is nowhere to be found in the computational model of perception.

And what about human perception? Apparently, the perceptual promiscuity of plants is alien to our highly specialized sense organs: the human eye is for seeing, the ear is for hearing, etc. The additional problem this raises is how to synthesize sense data in a perceptual presence that would match the unity of a perceived object. Kant is convinced that synthesis happens after the fact of receiving sense data and that it requires the cognitive intervention of the categories. Perception is the outcome of this combinatory effort, enabled by acts of thinking. Here, too, human perception is already in itself human thinking. But the sense data are themselves senseless and confused without the empty forms of reason molding them into coherent units. Kant writes: "Since, however, if representations reproduced one another

[6] Francisco Paco Calvo, "The Quest for Cognition in Plant Neurobiology." *Plant Signaling & Behavior* 2, no. (July/August 2007): 208–211.

without distinction, just as they fell together, there would in turn be no determinate connection but merely unruly heaps of them [*bloß regelose Haufen derselben*], and no cognition at all would arise, their reproduction must thus have a rule in accordance with which a representation enters into combination in the imagination with one representation rather than with any others" (*CPR* A121).[7] If Kant were to concede that plants had sensory and perceptual abilities, he would have ascribed to them the anarchic, unruly heaps of representations unfiltered through the categories.

The philosophical problem of human perception, which Kant emblematizes, is fictitious; it merely recaps what the mind-body and subject-object splits have cunningly introduced into our images of ourselves and the world. Certain kinds of art, neurological conditions, and altered states of consciousness give us a peek at synesthesia in human experience, but it remains an exceptional (and, perhaps, a disturbing) occurrence. Not just hallucinogenic plants, but all plants can teach us how to be polymorphously perverse in perceiving. To begin with, it is necessary to disrupt the functional-structural exclusivity of human organs and, thereafter, to verify their (our) vegetality. As we will see shortly, one can hear with the ear, as well as with the skin that picks up minute vibrations in the air, and finally with the entire body. Touch is equally dispersed through our corporeality, its organ—the skin. The perceiving and the perceived never come in isolated dyads, but in meaningful heard-seen-smelled (and so on) meshes. Perception is a complex middle which cannot be broken down into atomic components and from which the extremes of thinking and sensation can be removed with the same degree of success as a tree trunk can be chopped at two points along its extension in the hopes that this midsection would stay alive.

An articulation of the Parmenidian "thing" that is the same for thinking and for being, perception is, by the same token, a unique middle edge made of many edges blending into one another. Plant-perceiving is plant-thinking; human-perceiving is human-thinking—and plant-thinking in disguise. The multiple edges that intermingle in perception are those pertaining to the different senses, to cognition and sensation, and to the human and the plant.

Plant-thinking is neither conceptual nor pictorial. It is the non-theoretical contemplation of the world, a contemplation that does not keep itself aloof to the contemplated but is receptive and engaged, sensuous and interpretative. We may glean from it the idea that thinking is not only thinking (within the narrow register of abstract cognition and judgment) and that the dynamism of thought is indebted to this excess of thinking, this non-identity of thinking with itself in its interlacing with perception. Formulaically expressed: thinking $\neq$ thinking. That non-identity is the exact opposite of everything the algorithmic functions of contemporary scientific reductionism tell us. It is where the freedom of plants (and our own) takes refuge. If to think is to be free—not in the sense of a stale idealism but in the deeply

[7] References to Kant's *Critique of Pure Reason* appear in brackets, containing the pagination of the first (A) or second (B) editions. The English translation used is Immanuel Kant, *Critique of Pure Reason (The Cambridge Edition of the Works of Immanuel Kant)*, edited and translated by Paul Guyer and Allen W. Wood (Cambridge: Cambridge University Press, 1999).

materialist sense I have hinted at here—then plants are the most thinking and the freest of beings. Yet, for much of our history, we have been dealing with them without thinking (which is to say, without really perceiving them), betraying both plants and ourselves. Becoming human is learning to think with our sensing-perceiving-thinking bodies, with the world, and with the *aesthesis* of vegetal life.

Plant-Listening

There is no pedagogy without listening, without attentive listening to the other. Plant-listening is a particular, if highly unusual, instance of plant perception. As such, it follows the rules of plant-thinking that flies "in the face of the insanity of transcendent thought"[8] and thinks in the immanence of existence, with and as the other. Given that plant-thinking is, simultaneously, the thinking of the plants themselves, our thinking about plants, and the dynamic interaction between these two processes, plant-listening is how the plants themselves listen, our listening to them, and the resonance of their and our auditory attentions.

First: the listening of the plants themselves. Without anatomical parts corresponding to the ears, they are *all ears*, the vegetal body (or bodies: the proliferation of semi-autonomous growths within "one" plant) attuned to the minutest of vibrations above and below ground. To be all ears is to be exceptionally attentive, open to others and to the world, involved and concerned, non-indifferent. Plants pay attention to more than twenty environmental conditions, from humidity gradients to a wide spectrum of light wavelengths.[9] Why would they not be sensitive to sounds and subtler vibrations of the air or the earth?

From roots using sound to locate underground sources of water, through adaptive germination and growth responses to auditory stimuli, to leaf vibrations (that result from herbivore chewing) transformed into biochemical distress signals, plants are acoustically astute. But, perhaps, we should also hear the idiomatic expression *to be all ears* literally: the functional equivalent of animal and human ears is not localized in one organ; it is distributed throughout the plant, which listens with its entire body to its whole environment. Leaves, flowers, and roots are all of them, among other things, ears. Because vegetal ears do not have a circumscribed, proper site, they overlap with the plant's lungs and mouths, eyes, and skin. All ears, the plant is all eyes and all thinking, all vibrant and vibrating with sensory, perceptual, cognitive attention in every part.

Second: our listening to plants. This has been an unmitigated failure. With the exception of certain poetic insights and the budding, though still quite small, scientific field of plant bioacoustics, the consensus on the silence of vegetation has been

---

[8] Michael Marder, *Plant-Thinking: A Philosophy of Vegetal Life* (New York: Columbia University Press, 2013), 168.

[9] Refer to Anthony Trevawas, *Plant Behavior and Intelligence* (Oxford: Oxford University Press, 2014).

overwhelming. We take the fact that plants lack actual ears and an apparatus for generating sounds akin to ours or to that of animals who are similar to us for evidence of their muteness and deafness, sheltered from the very possibility of hearing. We treat them as ontologically disabled. Their *logos* without voice resonates for us with total silence. We, who without giving it a second thought subscribe to such a view, judge plants in advance, pre-judge them, and it is this prejudice that deafens *us*. For too long their hypersensitivity has slipped through the cracks of our crude perceptual and cognitive systems. In 2023, this failure of discernment is inexcusable.

Third: the resonance of the human and vegetal auditory attentions has been, consequently, fraught and doubly asymmetrical. While plants co-vibrate with us and with the rest of the world, we do not; while they are all ears, we have only two that, to add insult to injury, are too rudimentary to pick bioacoustic signals emanating from plants. Just as plant-thinking requires the painstaking work of reconfiguring, or even transfiguring, "the symbiotic relation" between human thought and vegetal existence,[10] so also the attunement of human listening to that of plants needs to be refined, finessed. How?

A key task is getting used to listening with our entire bodies, rather than exclusively with the ears. In fact, it is a matter of merely discovering something that is already there, as opposed to acquiring a new skill. It is a matter, then, of laying bare the repressed vegetality of our own bodies and psychic strata. Vibrations do not strike the ear alone; they affect and unsettle every material medium, though sentient surfaces are especially adept at registering them. The first sentient surface of our bodies is the skin, and it turns out that its vibrations convey information to the brain in a way similar to the auditory system.[11] The skin is probably our most vegetal organ, breathing through its pores like a leaf, sensing light without being an eye, hearing vibrations without being an ear. An organ as versatile, multifunctional, and exposed as those of plants.

The scope of what we listen to could also change following the lead of plants. Besides technologically altering the perceptual thresholds, within which we typically receive sounds, we would no longer seek speech and song, howling and chirping, hissing and roaring as the indicators of communication, accomplished exclusively by means of vocalization. Clicking frequencies, such as those maize roots emit, or insect buzzing in "buzz pollination" that triggers the release of pollen at an opportune time when the pollinator is around[12] have been heard (sometimes using sophisticated equipment that amplifies our bodily capacities) but not really listened to. The expanded scope of listening entails a transition from hearing to interpretatively attending to the sound waves and vibrations, which plants either

[10] Marder, *Plant-Thinking*, 10.

[11] Georg von Békésy, "Similarities between Hearing and Skin Sensations." *Psychological Review* 66, no. 1 (1959): 1–22; E. L. Mackevicius, M. D. Best, H. P. Saal, S. J. Bensmaia. "Millisecond Precision Spike Timing Shapes Tactile Perception." *Journal of Neuroscience* 32, no. 44 (2012): 15309.

[12] Monica Gagliano, "The Flowering of Plant Bioacoustics: How and Why." *Behavioral Ecology* 24, no. 4 (July–August 2013): 800–801.

produce or receive, as carriers of meaning. It invites a different approach to plants, not only as organic materials to be interpreted but as the interpreters of their world.

The resonance of distinct modes of listening deserves further analysis. It is the resonance of ears (or their equivalents), not of voices, which means that it is markedly silent. A vibrant silence permeates plant-listening, the silence within which speech begins to makes sense and to which it must return. Historically, the silences resonating in the listening of the plants themselves and in our listening to plants have been at loggerheads: one is pregnant with meaning, the other blocks the semantic flows emanating from everyone and everything not human. Sure, there have been (and there still are) cultures, groups, and individuals who do not fit my description and whose ears have been exquisitely receptive to the vegetal. Wonderful as it is, their existence does little to interfere with the tradition of plant-listening featuring the resonance of all and nothing, of vivacious openness and austere closure, of heedful excess and dearth of attention. Our goal should be not to carve out some exceptional niches of acoustic sensitivity in the general deafness to plants, but to turn the cultural tide of the ontological neglect of plants drowning them in human interests; not merely to prick up our ears and pick the previously inaudible signals, but to intervene in the interactive (or interpassive) dynamics of plant-listening.

Phenomenologically speaking, the promise of explorations in vegetal acoustics is that it expands the fields of *noesis* and *noema*, of the listening and of the listened-to. The difference between listening and hearing is crucial here. This difference has nothing to do with the fact that the former is attentive and saddled with the interpretation of sounds, whereas the latter is not. One may hear anything whatsoever, the noises coming from every direction; listening, however, always listens to the other. So, plant-listening, in the second sense of our listening *to* them, is *ipso facto* a method of treating plants as others, irreducible to the vague background of white noise or silence. Folded into the listened-to are the attended-to and the respected. And the multiplication of noematic layers does not end there: since the plant listens to (and listens in on) its entire environment, listening to the plant, we eavesdrop, with and through it, on the rest of the world. Through vegetal vibrations, we inch closer to the vibrancy of being.

Coda

These are but the beginnings of a vegetal environmental pedagogy that calls for a mutation in our thinking about thinking itself, about perception, and, especially, listening. We need to listen to plants, though not in order to interpret the "information" they might convey to us, but to be guided by them. We would do well if we refrained from imposing our communicative model on plants and, instead, heeded their ethical, aesthetic, and other kinds of lessons. Our fate and that of the non-human world depend on this.

This chapter has benefited from the contribution of the project PID2021-126611NB-I00 Socioecos. Building sustainable society: Mobilization, participation and management of socio-ecological practices, supported by the Spanish Ministry of Science and Innovation.

Bibliography

Calvo, Francisco Paco. 2007. The Quest for Cognition in Plant Neurobiology. *Plant Signaling & Behavior* 2 (4): 208–211.

Chamovitz, Daniel. 2012. *What a Plant Knows? A Field Guide to the Senses of Your Garden and Beyond*, 170–173. Oxford: Oneworld.

Dostoyevskiy, Fyodor. 1876a. *Civilization's Very Last Word.* Diaries, July–August. https://fedordostoevsky.ru/works/diary/1876/07-08/04/.

———. 1876b. *"The Earth and Children" [my translation].* In Diaries, July–August. https://fedordostoevsky.ru/works/diary/1876/07-08/13.

———. 1876c. *What Helps at Thermal Baths? Water or Good Manners?* In Diaries, July–August. https://fedordostoevsky.ru/works/diary/1876/07-08/04/.

Gagliano, Monica. 2013. The Flowering of Plant Bioacoustics: How and Why. *Behavioral Ecology* 24 (4): 800–801.

Kant, Immanuel. 1999. *Critique of Pure Reason*, Edited and translated by Paul Guyer and Allen W. Wood. Cambridge: Cambridge University Press.

Mackevicius, E.L., M.D. Best, H.P. Saal, and S.J. Bensmaia. 2012. Millisecond Precision Spike Timing Shapes Tactile Perception. *Journal of Neuroscience* 32 (44): 15309–15317.

Marder, Michael. 2013. *Plant-Thinking: A Philosophy of Vegetal Life*, 10–168. New York: Columbia University Press.

Trevawas, Anthony. 2014. *Plant Behavior and Intelligence*. Oxford: Oxford University Press.

von Békésy, Georg. 1959. Similarities between Hearing and Skin Sensations. *Psychological Review* 66 (1): 1–22.

Michael Marder is IKERBASQUE Research Professor in the Department of Philosophy at the University of the Basque Country (UPV-EHU), Vitoria-Gasteiz, Spain. His writings span the fields of ecological theory, phenomenology, and political thought. He is the author of numerous scientific articles and monographs, including *Plant-Thinking* (2013); *Phenomena—Critique—*Logos (2014); *The Philosopher's Plant* (2014); *Dust* (2016), *Energy Dreams* (2017), *Heidegger* (2018), *Political Categories* (2019), *Pyropolitics* (2015, 2020); *Dump Philosophy* (2020); *Hegel's Energy* (2021); *Green Mass* (2021), *Philosophy for Passengers* (2022), and *The Phoenix Complex* (2023), among others. For more information, consult his website michaelmarder.org

Chapter 10
Tasting, Breathing, Eating—Acknowledging Basic Embodied Experience as a Vital Source for Transformation

Donata Schoeller

Abstract In this paper I reflect how basic embodied responses and feelings can be an inspiration to get past dominant patterns of thinking that separate the human from the rest of the world. To make my point, I build on the innovative work of Corinne Pelluchon and Eugene Gendlin. Both demonstrate how pre-reflective abilities of the human organism become powerful philosophical entry-points that have transformative ethical and political implications. Some examples will be presented. The first part of this paper is inspired by Pelluchon's unfolding of the ethical implications of taste and the epistemological implications of compassion, or pity. The second part of the paper elaborates Gendlin's experiential and co-generative body-environment concept that, among other things, allows to understand hunger as the red thread between body, mind and meaning. On the basis of these approaches, I strive to exemplify the experiencing body as a vital driving force to feel, think and mean beyond powerful cultural constructions inherent in the catastrophic crisis face today.

Keywords Taste · Compassion/pity · Hunger · Body-environment · Meaning · Language · Pelluchon · Gendlin

Tasting

Every time we eat, we quite literally bite into the world. Food binds us to other beings, human and non-human, to circuits of production, to conditions of work and commerce, to materials both natural and fabricated, to means of transportation, to

D. Schoeller (✉)
Philosophy Department, University of Koblenz, Koblenz, Germany

Philosophy Department, University of Iceland, Reykjavik, Iceland
e-mail: schoeller@uni-koblenz.de

L. Škof et al. (eds.), *Elemental-Embodied Thinking for a New Era*,
Sophia Studies in Cross-cultural Philosophy of Traditions and Cultures 42,
https://doi.org/10.1007/978-3-031-42119-8_10

environmental footprints connected to the choice of what we eat. To eat therefore is in itself denying any separation between me and the world, nature and nurture. When what we bite into is a result of mass-production, we also bite into the suffering of the animals, into an exploitative economical system that has degraded animals to cost-efficient things, we bite into de-humanized working conditions of slaughter-houses, we bite into agriculturally caused emissions that harm the climate and that exploit the global south for the excessive cravings of the richer part of the word, we bite into gen-technology and bio-chemistry. With this emphasis the French philosopher Corinne Pelluchon makes us aware how the act of eating today has implications that go beyond the borderlines of disciplines, to name but a few: ethics, epistemology, economy, politics.

Obviously, the term "taste" engages the double play of meaning in which the tastelessness of fast-food corresponds the horrific tastelessness of the human abuse of animals in food production. Yet, Pelluchon's emphasis on taste is more subtle than this. For me, it's a noteworthy example of a basic pre-reflexive capacity to relate in a most differentiated way to the world. When we taste something, we are already relating to the world before we think and speak. Tasting something is a mode of knowing, discerning and differentiating before we deliberately know, discern, and differentiate. As such, taste is worthy of philosophical attention, beyond the aesthetical focus. It is a condition of knowing that does not figure in standard epistemologies or ontologies. Taste's central role is emphazised by "the connection that exists between its physiological and its cerebral dimensions, but also intellectual, cultural, even artistic dimensions."[1]

Emphasizing taste as an elementary act of relating to the world opens the scope of basic epistemological and ontological categories in an intriguing way. This goes hand in hand with raising awareness for all too obvious organismic capacities that are pre-rational, yet foundational for relating, orienting ourselves and making sense of the world. By its taste we know if something is "good" or (has gone) "bad." We get a sense of this not by knowing about it, by observing it, by analyzing it, or applying categories. Taste manifests an embodied understanding that comes with the elementary and closest possible contact of touching, licking, biting, smelling and incorporating what we eat. Tasting is an embodied understanding of the world as not separate.

To not cultivate the taste of nourishment and care for dignified ways to appreciate it—this is how Pelluchon continues to elaborate her elementary links between nourishment, epistemology and ethics—is an act of oblivion and disrespect of our multi-layered connection to the world. The loss of taste that went along with Corona, which happened after her book came out, seems almost symbolic to me in this respect. A symbol of a global culture that does not respect the nourishing web. Thus, raising awareness for the embodied language of taste implicitly breaks through an obliviousness that has sedimented right down to very basic forms of living. This

[1] Corinne Pelluchon, *Nourishment. A Philosophy of the Political Body,* translated by Justin Smith, (London: Bloomsbury Academic, 2019), 43.

silent act of appreciating the taste of things, of becoming aware of the web of non-human and human contributions in our nourishment, is political and ethical at the same time. It's a step out of a blind drift towards ever more efficiency: "The things from which I draw life are not given as objects or as utensils, even when I make use of them, but rather trace a horizon where utility and production are not primordial. This is why they are called 'nourishment.'"[2]

The disgusting violence of mass-meat production, plus the neo-liberal economical focus on quantity, on maximally cost-effective outputs and profits, have reduced the value of quality, not just of things, but of life itself. This is how I understand Pelluchon's point. Life and its quality of course cannot be captured by quantifiable parameters. Life's quality is an experience. Pelluchon's ontological phenomenology—in critical response to the Heideggerian concept of "Geworfenheit," thrownness into the world[3]—opens our minds to the basic goodness of being alive in the midst of a sustaining web of interactions without which we cannot live. In a family related way, Hartmut Rosa contributes the term resonance to get at a kind of experience that makes life worth living in a way that is not captured by a quantifiable focus on income, career, status, power.[4] Resonance implies receptivity in order to experience the embodied response to what is occurring when we eat, meet, work, study, learn, relate. The kind of bliss if things taste good, the feeling tones when being with a friend one trusts, the kind of silent joy during a conversation that opens horizons, the bodily feeling of freshness when hiking through a forest, the kind of awe flowing through the body when encountering a wild animal, the tender fulfillment when stabilizing a child that begins to walk, the excitement of learning with a teacher that sparkles with enthusiasm, these kinds of experiential responses can be considered as basic forms of embodied gratitude and appreciation for the human and non-human web which nourishes us.

Pitying

Of course, there is another qualitative side of such basic forms of embodied understanding and relating. It plays out in what we call pity. Nietzsche has devalued feeling pity (Mitleid) as something that weakens us, Pelluchon re-values it exactly because of that. "Mitleid" still has this connotation, and pity is also a term that has negative connotations as in sentences like: "I don't want your pity." The simple word "pity" emphasizes the ordinariness of such feeling capacity, even though one can certainly also use compassion as a translation. Pity or compassion jumps at us after having read, heard or seen suffering. It can ruin our mood, it can haunt us at night. Pelluchon writes:

---

[2] Pelluchon, *Nourishment*, 5f.

[3] Martin Heidegger, *Sein und Zeit* (Tübingen: Max Niemeyer, 1967), 175.

[4] Hartmut Rosa, *Resonanz: Eine Soziologie der Weltbeziehung* (Berlin: Suhrkamp, 2018).

> To experience pity is to be disturbed in one's flesh by someone else and to be dispossessed of oneself. This movement is prior to any representation. When experiencing pity I do not say 'I', for I am not a spectator external to the world. It is why those who suffer from the suffering of animals also suffer, at the same time, from the fact that human beings do so much harm. They do not accuse their fellow human beings, but for them, all reasoning, all justification, all cleavage, and the separation between the 'I' and the 'non-I' do not succeed in burying the initial identification with the suffering other. Moreover, they feel that the violence towards animals exercised by their fellow humans weighs on their own innocence.[5]

Having pity, or feeling compassion, is a pre-judgemental, and more importantly, is a pre-separational mode of knowing. It is a mode in which one cannot separate oneself from the suffering one witnesses or knows about, even if wanting to. Being touched by the suffering of another, weakens the importance of a present agenda, disturbs the course of the day, and even one's own sense of soundness. Pelluchon's differentiated way of revaluing this kind of feeling as an expression of primary non-separation re-establishes a ground of thinking from which it does not make sense anymore to consider these felt dimensions as only subjective, or "secondary", as philosophical traditions suggest. Yet, the degradation of the felt dimensions of thinking is not just a philosophical problem. It has deeply engrained our ways of life. To give but one example:

Again and again I am baffled, when being informed in short pace of one devastating news after the next, with not the slightest space for processing the embodied response, such as compassion, grief, disgust, hopelesness: the murder of two children, the explosion of a bomb killing many, the destruction of a stretch of ocean due to oil-licking, and the next thing follows without further ado, with the same tone of voice, whether it is sports or the weather, or the index. News today is like fast food. The speed to eat and not taste corresponds the short time to take in news and not feel. One more symptom of a culture's deadening effect of an elemental kind of connectivity that plays out on an embodied, responsive, resonant and silent level.

Today, a special effort is needed, an explicative and reflective achievement, to spell out an understanding conveyed by experiencing a world. We live it, we taste it, we breath it, we experience it in the embodied response to what happens. Not only is the seeming simplicity of evidence most challenging when it comes to think-with and make sense of it.[6] Tasting and pitying are a qualitative happening-together of existences, spheres and circumstances that common logic keeps separate. These embodied modes of understanding challenge conceptual implications that build on principal separations as the ground of thinking and making sense of the world.

---

[5] Pelluchon, *Nourishment*, 139.

[6] Donata Schoeller, "Towards a concept of a freedom to make sense," in Kevin Krytzka and Eric Severson, *The Philosophy and Psychology of Eugene Gendlin: Making Sense of Contemporary Experience* (London: Routledge, 2023).

Struggling with Language

Linguistic meaning is a storage of experiential ingredients from previous generations, life and culture-forms going back for centuries. Metaphorically spoken, language provides us with meals from the past, pre-cooked ingredients of meaning, from our own life-span, as well as from lifes of generations, that re-enter our organism when we use it. This is one way to understand the force of the language-battles we witness today. Yet, canceling or just changing the words is not enough and was never enough to re-new the fabric of meaning.

Whereas the power of language is unquestioned, the vastness of pre-verbal ingredients of meaning are still scarcely recognized. Meaning grows from enacting one's life, experiencing rhythms, relations, sufferings, and joys long before one can speak.[7] It is rooted in involvements that undermine the orderly distinction between things that come later, verbally. Meaning grows from childhood on by living with others within intricate situations,[8] and by interacting with various kinds of environments.[9] Meanings of a language, its expressions and ways to say something, take along generational stories and histories, entanglements of circumstances. Understanding verbal meaning comes about by complex constellations of active and passive learnings, that involve more-than-human participants etc.[10] This is why human bodies ever so often struggle to make sense of their world in the language that is available.[11] The complexity of events implies the complexity stored in language.[12]

Pelluchon suggests returning to the qualitative dimension of nourishment, taste and also hunger as a starting point for a transformative culture critique on the basis of rudimentary existentials of the human life. In a similar spirit, Eugene Gendlin's

---

[7] Alva Noë, *Out of our Heads: Why You Are Not Your Brain and Other Lessons from the Biology of Consciousness* (New York: Macmillan, 2013); Daniel N. Stern, *Forms of Vitality: Exploring Dynamic Experience in Psychology, the Arts, Psychotherapy, and Development* (Oxford: Oxford University Press, 2010).

[8] Wilhelm Dilthey, *Der Aufbau der geschichtlichen Welt in den Geisteswissenschaften* (Göttingen: Vandenhoeck und Ruprecht, 1992); Stanley Cavell, *The Claim of Reason: Wittgenstein, Skepticism, Morality, and Tragedy* (Oxford: Clarendon, 1979).

[9] John Dewey, *1938: Logic: The Theory of Inquiry* (Carbondale: Southern Illinois University Press, 1991); Eugene T. Gendlin, *A Process Model* (Chicago: Northwestern University Press, 2017).

[10] Karen Barad, *Meeting the Universe Halfway: Quantum Physics and the Entanglement of Matter and Meaning* (Durham: Duke University Press, 2007), and by the same author: "Intra-Actions: Interview with Adam Kleinman," *Mousse 34* (2012): 76–81; Donna Haraway, "Situated Knowledges: The Science Question in Feminism and the Privilege of Partial Perspective," *Feminist Studies* 14 (1988): 575–599, and by the same author: *Staying with the Trouble: Making Kin in the Chthulucene* (Durham: Duke University Press, 2016); Steven C. Hayes, Jason Luoma, and Robyn D. Walser, *Learning ACT: An Acceptance and Commitment Therapy Skills-Training Manual for Therapists* (Oakland: New Harbinger, 2007).

[11] Carl R. Rogers, *On Becoming a Person: A Therapist's View on Psychotherapy* (London: Constable, 2004).

[12] Donata Schoeller, *Close Talking: Erleben zur Sprache bringen* (Berlin: De Gruyter, 2019).

works with philosophically neglected kinds of knowing and understanding, manifested in breathing, walking, in hunger, in bodily behavior and feeling. In particular, his main work *A Process Model* is a "back to the basics," motivated to transform foundational epistemological and ontological approaches, to get us beyond implicit thinking patterns that have co-created the multiple crises we face today.

In Gendlin's work, a red thread becomes visible, a continuum from what we consider mere "bodily" interacting with the world, like eating, to high-level cognitive capacities of "the mind," like speaking, reading, using language. For instance, being "satisfied" after a good meal and after a formulation that works well has a noteworthy resemblance. People can become ill if they are not granted the conditions to say what they mean. Psycho-somatic harm is imposed on individuals when a political systems does not grant conditions to cultivate the capacity of saying what one feels, experiences and thinks.[13] In psychology, Alexithymia is a name to describe the complex symptoms that are connected if one has no words for emotions and feelings.[14] Acknowledging these phenomena is re-vitalizing the old intuition that we live from bread *and* meaning.

Thinking with the Body to Grasp Meaning

Language in terms of gains for coordinative action, as a means to build symbolic meta-stories in which successful group and mass-coordination became possible, is a line of thought elaborated by Ernst Cassirer,[15] and today by contemporary cognitive science. As illuminating as such an approach is to understand the evolutionary gain of something as complex as a language system, it does not offer clues for the kind of existential need for being able to express oneself. This need might even imply to break out of the meta-story that regulates the thinking and acting of a group. How can an individual be able to do so? Here the classical pragmatist and also enactive approaches seem more promising, by locating the roots of meaning within the ability of organisms to sustain their integrity while interacting and coping with environmental challenges. Gendlin's approach strengthens this line of thinking by opening an additional track. This track enables him to develop and understand the plasticity of meaning encoded in language.

[13] Hans Joachim Maaz, *Der Gefühlsstau: Psychogramm einer Gesellschaft* (München: Beck, 2019).

[14] John C. Nemiah, "Alexithymia and Psychosomatic Illness," *Journal of Continuing Education* 39 (1978): 25–37; Peter E. Sifneos, "The prevalence of 'alexithymic' characteristics in psychosomatic patients," *Psychotherapy and Psychosomatics* 22, no. 2 (1973): 255–262; Anna da Silve, Antonio Vasco, Jeanne Watson, "Alexithymia and Emotional Processing: a longitudinal mixed methods research," *Research in Psychotherapy Psychopathology Process and Outcome 21* (2018): 40-54.

[15] Ernst Cassirer, *Philosophie der symbolischen Formen* (Darmstadt: Wissenschaftliche Buchgesellschaft, 1964).

With only a few strokes, Gendlin's *A Process Model* dissolves the concept of a human body linked to the world only by sense perception. The traditional skepticism accompanying this approach goes hand in hand with hardly any awareness of the implicit richness of embodied connectivity that is in place before the philosopher's construction of epistemological or transcendental ties between subject and object.

In *A Process Model*, body is conceived *as* interaction with environment. John Dewey already warned about a thinking habit that might none-the-less assume body and environment as givens, that only then interact.[16] Gendlin's concept of the body undermines this strong habit, by anchoring the concept in the experience of the living body. Rob Parker emphasizes the experiential starting point of Gendlin's work: "[...] our chest expands as the air goes in, and contracts as the air goes out. But it's not as if the chest expands first and then the air goes in; chest expanding and air going in are the same event (same going on, same happening, same occurring)."[17].

Breathing undermines clear demarcation lines between body and world. Breathing means that oxygen enters the bloodstream-environment and goes all the way into the cells. Breathing impacts the way we think, feel and act. In such elemental engagement, where does the "body" end and the "environment" start, where does the contact between "body" and "environment" stop? Focusing on interactions that keep us alive, borderlines between body and environment are no easy question anymore. From any bone, paleontologists can derive not just the rest of the body but also the kind of environment and terrain in which the animal lived.[18]

Body-Environment—A Generative Process

Starting from lungs and air, me and other, environment and body, separates the living process according to nouns and grammatical structures one cannot get past anymore. When thinking from the experience of breathing, from walking, from hunger, from tasting, from the pre-reflective embodied response of feeling pity, the body challenges the mind to overcome bifurcations, and forge a new language. The body is not just "in" an environment. The body itself becomes recognizable as an environment for the process that keeps it alive. Gendlin writes: "The bloodstream is often called the environment of the cells it feeds. The many processes in the body have various parts of it for their environment. The skin-line is not the great divide. [...] The body **is** an environment **in which** body-process goes **on further**."[19]

Gendlin's sentences very often manifest of re-forging of the structure of language on the basis of experiential embodied understanding: "Just as my foot cannot

[16] Dewey, *Logic*.

[17] Robert Parker, "Introduction," in: Gendlin, *Process Model*, xii.

[18] See Gendlin, *Process Model*, 7.

[19] Gendlin, *Process Model*, 6.

be the walking kind of foot-pressure in water, we occur differently when we are the environment of each other. How you are when you affect me is already affected by me, and not by me as I usually am, but by me as I occur with you. We want to devise concepts to capture this exact aspect of 'inter- action first': what each is within an interaction is already affected by the other."[20]

There is an unsurmountable difference between an observer perspective on body and environment, implying the separation of subject and object, and an interaction-first approach that cannot separate between the two. In *A Process-Model*, Gendlin refers to body-environment as one term, indicating the interactional process, thus raising awareness for the generative dimension implicit in body-environment.[21] The body-process relies on an environment that evolves through the process: "The life process […] makes itself an environment in which it then goes on further."[22]

Human bodies are of course just one instance of this overarching principle. It applies to animals and plants, obviously also to eco-systems. Culture can also be understood as a structured environment generated by living processes, which have come to imply and go on in further in this 'home-grown' environment. Understanding the environment generativing dimension of living process bridges the nature-nurture, nature-culture divide. It's a training to think-with the closest kind of knit: living process and environment is but one mutually sustaining happening. This is also the case, when a human way of living generates and sustains systems that saw the very branch of the more-than-human living on which they rely.

The interactive and generative connection of living-process and environment helps to understand why in a living world, things can change drastically and at the same time why change is so difficult: the environmental structures have grown through the living processes which sustain them, and which again imply these very structures. For example,

> the ground you walk on is part of how your feet are built, and not only your feet. The muscles up into your thighs, your posture and the whole balance of all your organs already include the ground against which you press when you walk. The sensations and internal feel-quality of all your organs and muscles comprehend how you walk on solid ground. Your walking-pattern is implied in the feel of your whole body.[23]

This approach is a serious lesson to think within responsive complexity, which applies on a micro- as well as on a macro perspective. A certain habitual posture has imprinted a body in a way that implies that posture, which implies conditions of gravity as well as cultural norms of behaviour, as well as the way all other "parts" of the body are, including the feeling of being in that posture. A family, for instance, manifests the same principle: It has generated a homegrown environment that implies a way (of dysfunctional or functional) living together, which again implies

[20] Ibid., 31.

[21] Neil Dunaetz and Donata Schoeller, "Thinking Emergence as Interaffecting: Approaching and Contextualizing Eugene Gendlin's Process Model." *Mind and Matter* 16, (2018).

[22] Ibid., 7.

[23] Eugene T. Gendlin, *Let Your Body Interpret Your Dreams* (Wilmette: Chiron, 1986), 143.

histories, cultural settings, values, sedimented in the possibilities and constraints of communicative patterns, behaviors, ways the family-members' bodies feel and act when being together, thereby sustaining the very structure. Obviously, same applies to what we call political, economic, and cultural systems.

In his article in *Arts of Living on a Damaged Planet*, Andrew Mathews opens one's eyes to the complexity of interactional traces of the living processes in smallest details "tangible through the form of ancient chestnut stools. Centuries of grating, cultivation, trade, taxation, and disease are inscribed onto their structure and shape. The landscape emerges from [...] entanglements: the many histories of life and death that have made these trees, this place."[24] Things that seemingly do not belong together according to logical categories and disciplinary classifications, do indeed belong together to understand the very make up things. What Elaine Gan et al. call "ghosts from the past," a metaphoric reminder of the entangled tracks of "more-than-human histories through which ecologies are made and unmade,"[25] is at the heart of Gendlin's interactional model. A stool is as good a manifestation of this model, as is a – damaged or a thriving – planet. When it comes to living systems, hierarchies of high-level and low-level processes miss an important point: "Even a very tiny process in some microscopic location implies (is also the implying of) the larger processes of which it is a part. The various processes each imply the continuance of the others *as the whole process*, and thereby also its own continuation.[26]

Spelling out an interactional approach as carefully as is done in *A Process Model*, provides strong conceptual handles to get a grip on the important difference of a transformation that happens through interactional living, and a disruptive change from outside the interwoven fabric of sustaining webs of living. Disruptive change, in our time, can be considered as a result of a way of thinking that produces a "single story," a concept I borrow from Chimamanda Adichie.[27] Paraphrasing the term here, I mean to reference anthropocentric thinking as a single story that has made itself increasingly untouchable by the multi-dimensional stories of a living world in which our story is embedded. This single story has fostered action- and thinking-patterns that are neither responsive nor responsible for the more-than-human, driven by narrow and shortsighted perspectives of human success. Greedy and manipulative exploitations combined with vast technological leverage have thus created a human culture in which "living arrangements that took millions of years to put into place are being undone in the blink of an eye."[28]

---

[24] Andrew S. Mathews, "Ghostly Forms and Forest Histories," in *Arts of Living on a Damaged Planet: Ghosts and Monsters of the Anthropocene*, ed. Elaine Gan, Anna Lowenhaupt Tsing, Heather Anne Swanson, and Nils Bubandt (Minneapolis: University of Minnesota Press, 2017), G145–G156.

[25] Gan et al., *Arts of Living*, G1.

[26] Gendlin, *A Process Model*, 39.

[27] Chimamanda Adichie, "Die Gefahr einer einzigen Geschichte," filmed 7.10.2009, TED video, 19:15, https://www.youtube.com/watch?v=D9Ihs241zeg.

[28] Gan et al., *Arts of Living*, G1.

Experiencing Language—Re-freshing Meaning Beyond De-construction

Yet, how do we get out of a story that has become a hostile environment for most of the more-than-human lifes, while deeply imprinted how we think and speak about the "rest" of the world, about elements, animals, and plants as of "resources" to be used, consumed, and manipulated according to human needs? Use of language and patterns of thinking are interconnected. Much of philosophical, especially feminist and environmental thinking is characterized by critizing implicit assumptions and patterns inherent in the way we think and speak, and more so, by the effort to overcome them.[29] I believe, attending to *what it is like* to experience the most basic acts is crucial for this transformative endeavour. Becoming aware of what it is like to breath, walk, taste or pity is a thread to ways of being that are ongoingly interaffected and interaffecting. Gendlin has created the verb "eveving" (everything affecting everything) to indicate the intricacy of such interconnected embodied experience.[30] For this reason, attending to the experiential dimension of anything can be a powerful teacher out of reductive patterns and conceptual cuts of thinking.

Even the experience of how one uses language can offer itself as such a path. For instance, one can sense that an habituated language-use is "off" in some ways. Ralph Waldo Emerson gave a canonical description of this kind of experience, finding precise words for the "chagrin" in face of an alienating language-game:

> Do I not know beforehand that not possibly can he say a new and spontaneous word? [...] This conformity makes them not false in a few particulars, authors of a few lies, but false in all particulars. Their every truth is not quite true. Their two is not the real two, their four not the real four; so that every word they say chagrins us, and we know not where to begin to set them right.[31]

Further on, Emerson concludes that this kind of language encloses us in "prison-uniforms" of different parties. How strikingly to date these words sound, when we are being forced into polarizing games that make less and less sense. Yet, in these very lines Emerson subtly opens the prison-uniform he characterizes. And he does so not by arguing philosophical positions about language. Emerson speaks of an experience. He gives simple and direct words to the experience of pre-verbal, deeply embodied frustration of a language of his age and time. Even though it seems logically impossible to get beyond a language that has been the conforming and restraining environment of making sense, the experience of it renders cracks into the "prison-uniform."

Understanding how this is impossible is a philosophical project with a practical aim: to cultivate skillfull attention to the experiential dimension in order to open the powerful boundaries of a discourse. While this seems to resemble the

---

[29] Karen Barad, *Meeting the Universe Halfway*, chapter 4.

[30] Gendlin, *Process Model*, 43.

[31] Ralph W. Emerson, *Self-Reliance and Other Essays*, ed. Stanley Appelbaum (New York: Dover, 1993), 23.

deconstructivist's project, its materials are not only "texts" or "discourses." It works with the experiential texture of the feeling body to get at the transformative cracks to leverage openings in solid patterns of thinking and speaking. Thinking meaning in terms of a generated environment not unlike a body, is a way to understand the transformative possibilities involved. Within the scope of this paper, allow me to sketch a few perspectives which obviously need more work to be fully fledged out.

Meaning, just like a body, has been and is evolving interactionally. Meaning's soil, from which it grows and changes, is what we call situations. Situations for their part cannot be disconnected from a use of language implicitly and explicitly structuring them. This Wittgensteinian lesson is a benchmark behind which an embodied approach does not fall. Rather, an embodied approach thickens this understanding by adding the vast dimension of pre-verbal meaning situatedness involves. The lived experiences of life forms, histories and cultural patterns are implied right down into the depth of visceral and breathing patterns, of feeling and thinking habit.[32] Acknowledging the embodied dimension of meaning implies acknowledging its deep roots. These roots are unseparable from the body's own past, which is unseparable from the body's presence: "The past **is** the body. It carries the scars of my childhood cuts. It has the hangover from last night. It has [is, ...] my past experiences and those of the race and the species."[33]

Explicating the experiences of last night, of our childhood cuts, of our race or gender, is transformative in a way that cannot be described if we divide between language (belonging to the mind) and body. The experience of the present changes in an embodied way, when the past is verbally touched in certain ways. Re-traumatizing or healing are but two poles of the power of such touching with language.

Human bodies can paradoxically begin to transform powerful, disconnecting and dysfunctional systems that have sedimented down to the way they behave, speak, think, feel and breath—merely by attending in certain ways to the constricting and devastating experiences involved. Articulating the actual experiencing generated by the systems humans create can be an important first step out of the patterns holding these systems in place. John Dewey and Hannah Arendt, each in their own way, recognize the political and educational potential of facing actual experience as a starting point for genuine thinking. Finding words for what has been experienced, generates new understandings, the release of frozen feelings, novel resources of empathy or anger, fresh experiences that would otherwise be impossible. Attending to experience, and finding words for it, is a living process. As such, it is environmentally generative. It generates an understanding that is not just intellectual, but what I call an "embodied context".[34]

32 Gendlin, *Process Model*, Chapter VII.

33 Ibid., 61.

34 Schoeller, *Close Talking*.

Hunger for Transformation

Life process is "temporally organized," but here this does not mean only that someone notices hunger coming before eating. It means rather that hunger **is the implying of** eating. And eating? There is that special relation again: if hunger is the implying of eating, then eating is the '.....' of hunger. The term we want is implicit in the '.....' and when we get the term it will do to our '.....' what eating does to hunger. We can try out saying that eating satisfies hunger, that it carries out what hunger implies, that eating carries the hunger into some sort of occurring. Hunger **is** the implying of eating (the 'need' for food we say, making a noun out of this implying). Then eating is the satisfaction (another noun). The nouns make separate bits out of the process. But actually it's no fun eating if you're not hungry while eating.[35]

The short passage of Gendlin intriguingly indicates another read thread between seemingly elemental experiences and seemingly "higher standing" acts of the mind, like describing experience with language. This very example preforms what it says and thus shows that describing experience is not identical to constructing or representing something. Searching for a word to capture what it is like to eat into hunger is not an easy thing at all. One encounters a " ....," a gap, something that needs to be filled to satisfy, to carry that basic experience forward into language such that it is not alienated, simplified, left hanging or even distorted.

The passage indicates that hunger, as the most elemental experience, functions like a bridge between body and mind, forging new pathways with an existential and irresistible power. In an analogous way, an elemental feeling of hunger for transformation is a driving force, just like physical hunger. Like the fly that does re-iterating buzzes to get out of the closed window, the pre-verbal, and pre-separated embodied understanding of the scope of a crisis enacts a space from which novel movements can emerge.

This kind of hunger for change is embodied in the movement of a generation, which we call "Fridays to future," including the young pioneers of vegetarianism and veganism. The phenomenon of its power is an example for the energy that is unleashed when the understanding of what is missing in the human world is embodied in ways that can be felt like a hunger. The movement does not keep the sense of a crisis at arms length. It bears to feel the scope of it. It does not induce artificial boundaries to protects one's way of life, despite the non-separated sense of compassion with animals captured and slaughtered in mass-production. It acts upon an experience of non-separation, that is "reaching a truth that is older than thought, and to which the sciences, at least since Darwin and the beginning of research on animal ethology, confer their legitimacy."[36] By not numbing themselves from feeling the hurt in the more-than-human world, this generation cannot but crave for change a hungry body desparate for food. Letting the crisis come close in this way has generated unprecedented initiatives. Understanding the source of the power of her own

[35] Gendlin, *Process Model*, 10.

[36] Pelluchon, *Nourishment*, 140.

movement, Greta Thunberg begs grown ups to take on the burden of the close-up experience of a crisis in order to get going: "I want you to panic."[37]

An elemental hunger for change is the driving force for action. It is also the driving force to develop new ingredients of meaning that do not disconnect the human body from other living bodies, yet to "carry the living process further"[38]—in ways radically transcending anthropocentricity.

This hopeful picture obviously has a crux. An anthropocentric exploitation and destruction of the more-than-human world goes hand in hand with a habituated neglect of experience, the felt dimensions of thinking, the embodied dimensions of meaning. Environments co-generated by ever accelerating production-systems and consumerism, including the speed and amount of work cultivated within educational institutions—all this sucks away time and courage to stay engaged with the felt hunger in transformative ways. A consumerist culture sustaining a disconnection from elemental ways of experiencing and feeling fuels outbursts of anger and desperation, instead of collective processes that allow to hold and process the experience of a crisis, in order to find new steps, new ways to speak to each other, new ways to care for and bear such experiencing. Therefore, far from being an environment in which the human and more-than-human entanglements can unfold in novel ways, language today is conceived like the running of a program. We are offered AI chats that can do the languaging for us. The exhausted mind that exploits the world for the sake of energy, lacks its own energy to change the status quo.

Listening to feelings of fatigue, ecological grief and anxiety is thus equivalent to listening to a muted embodied agency—which *is* interaction with environment. In an age and time called the Anthropocene, taking these embodied responses seriously, is a key to unlock stuck patterns—on a personal and collective level.

Bibliography

Adichie, Chimamanda. *Die Gefahr einer einzigen Geschichte*. Filmed October 7, 2009. TED video, 19:15. https://www.youtube.com/watch?v=D9Ihs241zeg. Accessed 17 May 2023.

Arendt, Hannah. 1979. *Vom Leben des Geistes*. München: Piper.

Barad, Karen. 2007. *Meeting the Universe Halfway: Quantum Physics and the Entanglement of Matter and Meaning*. Durham: Duke University Press.

———. 2012. Intra-Actions: Interview with Adam Kleinman. *Mousse* 34: 76–81.

Cassirer, Ernst. 1964. *Philosophie der symbolischen Formen*. Darmstadt: Wissenschaftliche Buchgesellschaft.

Cavell, Stanley. 1979. *The Claim of Reason: Wittgenstein, Skepticism, Morality, and Tragedy*. Oxford: Clarendon.

[37] WELT Netzreporter, "I want you to panic" – Greta Thunberg hat eine klare Botschaft an die Mächtigen., YouTube video, posted January 25, 2019, accessed May 17, 2023, https://www.youtube.com/watch?v=ggKPH746dyw.

[38] Eugene T. Gendlin, "The New Phenomenology of Carrying Forward," *Continental Philosophy Review* 371 (2004): 127–151.

Da Silve, Anna, Antonio Vasco, and Jeanne Watson. 2018. Alexithymia and Emotional Processing: a longitudinal mixed methods research. *Research in Psychotherapy Psychopathology Process and Outcome* 21: 40–54.

Dewey, John. 1991 [1938]. *Logic: The Theory of Inquiry*. Carbondale: Southern Illinois University Press.

Dilthey, Wilhelm. *Der Aufbau der geschichtlichen Welt in den Geisteswissenschaften*. Göttingen: Vandenhoeck und Ruprecht.

Dunaetz, Neil, and Donata Schoeller. 2018. Thinking Emergence as Interaffecting: Approaching and Contextualizing Eugene Gendlin's Process Model. *Mind and Matter* 16.

Emerson, Ralph W. 1993. *Self-Reliance and Other Essays*, edited by Stanley Appelbaum, 23. New York: Dover.

Gendlin, Eugene T. 2017 [1992]. *A Process Model*. Chicago: Northwestern University Press.

———. 2004. The New Phenomenology of Carrying Forward. *Continental Philosophy Review* 371: 127–151.

Haraway, Donna. 1988. Situated Knowledges: The Science Question in Feminism and the Privilege of Partial Perspective. *Feminist Studies* 14: 575–599.

———. 2016. *Staying with the Trouble: Making Kin in the Chthulucene*. Durham: Duke University Press.

Harari, Yuval N. 2011. *Sapiens: A Brief History of Humankind*. London: Vintage Books.

Hayes, Steven C., Jason Luoma, and Robyn D. Walser. 2007. *Learning ACT: An Acceptance and Commitment Therapy Skills-Training Manual for Therapists*. Oakland: New Harbinger.

Heidegger, Martin.1967. *Sein und Zeit*, 175. Tübingen: Max Niemeyer.

Maaz, Hans Joachim. 2019. *Der Gefühlsstau: Psychogramm einer Gesellschaft*. München: Beck.

Mathews, Andrew S. 2017. Ghostly Forms and Forest Histories. In *Arts of Living on a Damaged Planet: Ghosts and Monsters of the Anthropocene*, ed. Elaine Gan, Anna Lowenhaupt Tsing, Heather Anne Swanson, and Nils Bubandt, G145–G156. Minneapolis: University of Minnesota Press.

Nemiah, John C. 1978. Alexithymia and Psychosomatic Illness. *Journal of Continuing Education* 39: 25–37.

Noë, Alva. 2013. *Out of our Heads: Why You Are Not Your Brain and Other Lessons from the Biology of Consciousness*. New York: Macmillan.

Parker, Robert. 2017 [1992]. Introduction. In *A Process Model* by Eugene T. Gendlin. Chicago: Northwestern University Press.

Pelluchon, Corine. 2019. *Nourishment: A Philosophy of the Political Body*. Trans. Justin Smith, 43. London: Bloomsbury Academic.

Petitmengin, Claire. 2021. Anchoring in Lived Experience as an Act of Resistance. *Constructivist Foundations* 162: 172–181.

Rogers, Carl R. 2004. *On Becoming a Person: A Therapist's View on Psychotherapy*. London: Constable.

Rosa, Hartmut. 2018. *Resonanz: Eine Soziologie der Weltbeziehung*. Berlin: Suhrkamp.

Schoeller, Donata. 2019. *Close Talking: Erleben zur Sprache bringen*. Berlin: De Gruyter.

———. 2023. Towards a concept of a freedom to make sense. In *The Philosophy and Psychology of Eugene Gendlin: Making Sense of Contemporary Experience*, ed. Kevin Krytzka and Eric Severson. London: Routledge.

Sifneos, Peter E. 1973. The prevalence of 'alexithymic' characteristics in psychosomatic patients. *Psychotherapy and Psychosomatics* 22 (2): 255–262.

Stern, Daniel N. 2010. *Forms of Vitality: Exploring Dynamic Experience in Psychology, the Arts, Psychotherapy, and Development*. Oxford: Oxford University Press.

WELT Netzreporter. "I want you to panic" – Greta Thunberg hat eine klare Botschaft an die Mächtigen. YouTube video. Posted January 25, 2019. https://www.youtube.com/watch?v=ggKPH746dyw. Accessed 17 May 2023.

Donata Schoeller is a research professor of philosophy at the University of Iceland and associate professor at the University of Koblenz. She is the co-initiator of the international research project Embodied Critical thinking (ect.hi.is), and the Academic Director of the Erasmus+ Strategic Partnership in Higher Education's program Training Embodied Critical Thinking (trainingect.com). She is board member of Mind and Life Europe. Her main research areas are embodied theories of meaning, process philosophy, enactivism and classical pragmatism, as well as Medieval and renaissance philosophy, especially Meister Eckhart and Jakob Böhme. Among her recent publications are *Practicing Embodied Thinking in Research and Learning*, ed. with Sigridur Thorgeirsdottir and Greg Walkerden (London: Routledge, 2023), *Close Talking: Erleben zu Sprache bringen,* (Berlin: De Gruyter, 2019), *Saying What We Mean*, ed. with Ed Casey, (Chicago: Northwestern University Press, 2017), and *Thinking Thinking,* ed. with Vera Saller, (Freiburg: Alber Verlag, 2016), *Enthöhter Gott—Vertiefter Mensch. Zur Bedeutung der Demut bei Meister Eckhart, Jakob Böhme und Friedrich Nietzsche*, Freiburg: Alber 2009 2nd edition).

Chapter 11
Dancing with the Stars: Learning to Move like the Earthlings We Are

Emil Månsson

> *I adhere to this world which is not me as closely as to myself, in a sense it is only the prolongation of my body … As Bergson said in Les Deux Sources: my body extends unto the stars.*
>
> – Maurice Merleau-Ponty, *The Visible and the Invisible*
>
> *Nature is not to be controlled – indeed we control nothing – and when something results according to our design, it is because we have danced with nature instead of opposed it.*
>
> – Humberto R. Maturana & Gerda Verden-Zöller, *The Origin of Humanness in the Biology of Love*

Abstract If behind our ecological destruction lies a worldview of human exceptionalism and exemptionalism, at surface level our actual living—"our" referring to the civilizational majority and our coexistence—is characterized not by accumulation and acceleration, but, I will claim, by a trembling standstill. We think we are steadily evolving and growing, but in reality *we are the sedentary ones.* We grind and grind but never get anywhere. It's not that we don't move but that our movement is rigid and nervous. We have grown pandemically impaired, strangely immovable in spite of our incessant rush. How can we begin to relax this retrograde movement? No simple solution suffices, so I explore not a singular answer but an approach: dancing as a loving affirmation of an ontological intimacy that is endemic to the very fabric of our planet and our own embeddedness within a wider ecological self. How might a dancing cure reanimate and reconnect us—with nature, with the places we inhabit, with each other, with nonhuman people, with ourselves—and remind us of our responsibilities towards all Earthlings?

Keywords Dancing · Rhythm · Attunement · Connectedness · Loving · Resonance · Embodiment · Awkwardness

E. Månsson (✉)
Department of Philosophy, University of Iceland, Reykjavik, Iceland
e-mail: emilmaansson@pm.me

L. Škof et al. (eds.), *Elemental-Embodied Thinking for a New Era*, Sophia Studies in Cross-cultural Philosophy of Traditions and Cultures 42,
https://doi.org/10.1007/978-3-031-42119-8_11

I am not much for dancing. I tend to keep half an eye on the floor (*what am I doing here?*) and half an eye on the room (*what are they thinking?*), while a second full eye has somehow dislodged itself from my body and is sceptically surveying the situation from above (*this is not ending well*). There are exceptions to this—I can be perfectly uninhibited rocking out with my wife and kids at home, or under the influence—but outside the family or an altered state of mind, I am just too self-conscious, and cannot take myself or the situation seriously.

Something similar is true of embodied practices in philosophy. They plainly do not sit well with me. They feel forced, awkward, fake even, and so take a lot of effort. Which begs the question: How did I end up doing my PhD in a project on embodied critical thinking (ECT) in the first place? And why is it relevant here?

First, I want to acknowledge the awkwardness. Whatever your inclinations are coming into any new practice, it's alright. A lot of us are not natural born dancers, or we think we are not, but— and this is my second point—that doesn't mean we get to sit on our hands. The world is dying and it's time we all get out of our comfort zones to save it; and to save ourselves, in a literal and a moral sense alike.

Now, this "coming out" is made considerably simpler by the realization that we are already out, that we are already "in the out." Beyond my comfort zone lies a no less fragile and a lot more vital zone: the "critical zone" we all live in *and create* together.[1] So, rather than a question of coming out, it's a matter of the comfort zone disintegrating as we real-ize, i.e., make real to ourselves, that we are part of Earth's biofilm, indeed of its very being, which we are sorrily self-harming. That we are *critically embodied* with/in an epic ensemble of terrestrial e-co-existence, and always were. To imitate: as we come *back to the earth itself*, or back *down to earth*, we begin to see—in a wider sense of understanding via experiencing—that *we have never been comfortable*.[2] Which is another way of saying...

[1]The scientific concept of Earth's "Critical Zone" stems from the National Research Council's 2001 publication *Basic Research Opportunities in Earth Science*, designating "the heterogeneous, near-surface environment in which complex interactions involving rock, soil, water, air, and living organisms regulate the natural habitat and determine the availability of life-sustaining resources" (Washington, DC: The National Academies Press), 2. On its adopted application in and for contemporary eco-theory, see Bruno Latour and Peter Weibel, eds., *Critical Zones. The Science and Politics of Landing on Earth* (Cambridge, MA: MIT Press, 2020).

[2]See Charles S. Brown and Ted Toadvine, eds., *Eco-Phenomenology. Back to the Earth Itself* (Albany: SUNY Press, 2003), a pun on the Husserlian dictum "back to the things themselves"; Bruno Latour, *Down to Earth. Politics in the New Climatic Regime* (Cambridge: Polity Press, 2018); and Bruno Latour, *We Have Never Been Modern* (Cambridge, MA: Harvard University Press, 1993).

On Earth We're Always Awkward

Nothing hurts and heals quite like the awareness that what you thought you were you really aren't. The universal—indeed, pandemic—spread of awkwardness is the theme of Bruno Latour's Covid-19 book, *After Lockdown: A Metamorphosis*, which would also be his last. In it, he casts Kafka's character, Gregor Samsa, as a model for how we all find ourselves (wake up) corporally transformed in the ecological, coronavirus-exacerbated climatic regime (his term for the Anthropocene) of today:

> I still remember how, before, I could move around innocently taking my body with me. Now I feel like I have to make an effort and haul along at my back a long trail of CO_2 that won't let me buy a plane ticket and take off, and that now hampers my every movement, to the point where I hardly dare tap at my keyboard for fear of causing ice to melt somewhere far away. But it's been worse since January [of 2020] because, on top of that, I now project in front of me…a cloud of aerosols…capable of killing my neighbors…[3]

The climate and Covid-19 crises together highlight an inescapable aspect of modern existence. We are no longer, if we ever were, just *there*. We are widespread. On the upside, it means the essentialist dream of a "pure" *Dasein* (there-Being) need no longer be entertained, as the very place-ness or place-less-ness of our being is so dramatically called into question. But it also means our own embodiment is called into question. We find ourselves, suddenly or not-so-suddenly, out of place in a dual sense: we do not fit because *we* are made monstrous, innocuous no more, *and* because the world is, its classical elements of earth, air, water, and fire acting out—performing and misbehaving—impossible weather events with a frequency that has "in no time," we say, instituted a new normal that is anything but.

In 2003, the philosopher Glenn Albrecht coined the neologism "solastalgia" to describe the "homesickness you have at home," when environmental disintegration distorts a land you once knew and loved, from an amalgamation of the words solace, desolation, and nostalgia. Handy as this term is, I was surprised to learn that Albrecht misses or omits how *sol* is also a Proto-Indo-European root meaning "whole, well-kept, healthy" (as found in solid, solidarity, save, i.e., to deliver from danger, from *salvare*: "making safe, secure"), and that it means soil, earth, and ground in French. Which are helpful secondary meanings—in fact I'd say they are primary—for pointing to the *loss of wholeness* both of the land itself and in terms of hitherto healthy relations between a person/persons, or people/peoples, and a place; as well as to the *loss of solace* that comes (came) from an intimate connection to land, as in, a particular earth-ground—*sol*, akin to what in Australia is known as "country"[4] —as that

[3] Bruno Latour 2021, *After Lockdown* (Cambridge: Polity Press, 2021), 2–3.

[4] Deborah Bird Rose explains, "Country is not a generalised or undifferentiated type of place, such as one might indicate with terms like 'spending a day in the country' or 'going up the country.' Rather, country is a living entity with a yesterday, today and tomorrow, with a consciousness, and a will toward life. Because of this richness, country is home, and peace; nourishment for body, mind and spirit; heart's ease." See, *Nourishing Terrains. Australian Aboriginal Views of Landscape and Wilderness* (Canberra: Australian Heritage Commission, 1996), 7. For a similar use of "land," consult Aldo Leopold's so-called land ethic, e.g., "We can be ethical only in relation to something

land is altered beyond recognition, even, to use a politically-fascistically charged term, as it's being *disappeared.* There are no stable comfort zones as we enter, like coming to inside the Matrix, "out inside"[5] the critical zone of this dying world, meaning both: (1) the dying of the natural world, (2) the dying of the social world, and (3) the dying community, that is, all of us living through and with the first two senses of the phrase (though we're not of course living the dying equally; to paraphrase George Orwell, everybody dies but some die more than others, so, to say we are all awkward is not to suggest we are equally awkward, or equally dis-abled;[6] in a similar vein, let me also stress that "the world" is never one but is always unfathomably rich, even as it is ruined and impoverished beyond repair).

Solastalgia, then, entails strong original feelings of belonging and a sense of rootedness, and a proportionally great sense of grief and longing for that bygone connection-to-home once it is lost. As such, it speaks to a deepfelt dearth of what I will call "ontological intimacy" after Kym Maclaren—though she uses the term strictly for interhuman relations—or "ecological intimacy" after Timothy Morton.[7] This, often tragically stifled, sense of onto-ecological intimacy is, I believe, the emotional undercurrent that stimulates, say, Indigenous land claims and the rights-of-nature movement. Which makes sense considering the term solastalgia was conceived precisely to encapsulate the collective experience of autochthonous peoples whose primeval lands are destroyed by an encroaching civilization. In recent years, however, the term has gained currency in the centres of civilization too, as it evidently resonates with a lot of people (us) there (here) also. There's also nothing strange about this, when we consider that most of us so-called Moderns are in a sense indigenous in our own way. Albeit not to place, to a land or country, but to a

we can see, feel, understand, love, or otherwise have faith in." Leopold, *A Sand County Almanac. And Sketches Here and There* (Oxford University Press, 1987[1949]), 214.

[5] Latour writes, "Where the appeal of paradise prevented people from acting, the impossibility of realising paradise on earth has ended up paralysing all forms of action 'to get out' [...] But *what* do we need 'to get out' of? Our paradoxical answer, the answer of the locked-down, of the earthbound, is that we need to get completely out of this not-so-material matter. But: to go where? For pity's sake! *To go back home, to the place where you are, the place you've never left* [...] Revisiting this story requires of us, it's true, the suppleness of contortionists: to get out of it, we need to get out of the idea of getting 'out', and so we need to decide to *stay* and even to *go out inside*!" Latour, *After Lockdown*, 52–53.

[6] Waking up awkward, in Latour's Kafkaesque sense, lends itself nicely to commentary from disability and crip studies. As we aka *I*—able-bodied *normie*, poster boy of *the normate*—find ourselves crippled with/in a new normal, should we not look to and learn from those most accustomed to mis-fitting? See, for example, Robert McRuer, "Compulsory Able-Bodiedness," and Rosemarie Garland-Thomson, "Misfitting," both in Gail Weiss et al., eds., *50 Concepts for a Critical Phenomenology* (Evanston: Northwestern University Press, 2020), 61–68, 225–30. That study, though, isn't fit for here.

[7] See Kym Maclaren, "Intimacy and embodiment: An introduction," *Emotion, Space and Society* 13 (2014): 55–65; and Kym Maclaren, "Intimacy as Transgression and the Problem of Freedom," *Puncta: Journal for Critical Phenomenology* 1 (2018): 18–40. And see Timothy Morton, "Attune," in *Veer Ecology. A Companion for Environmental Thinking*, eds. Jeffrey Jerome Cohen and Lowell Duckert (Minneapolis: University of Minnesota Press, 2017), 154.

way of life: the imperial mode of living.[8] What characterizes this way or mode? Elsewhere, I have described how a specific perceptual mode which begins and ends with seeing and thus taking (grabbing) the world as essentially dead matter/material is the motor of an assault mentality that ultimately sanctions and legitimizes rampant resourcism and unrestrained economism; and how this mode of living and perceiving can only sustain itself by cannibalizing—we call it relying on—extra-imperial others and elsewheres, resulting in the equally heartbreaking and incriminating reality of our ceaseless destruction of our own habitat, what many of us regard as outright murder, i.e., ecocide.[9]

Here I will explore a different but related conceptualization. One that is corporeal rather than perceptual. Recall the epigraph in which Humberto Maturana and Gerda Verden-Zöller write that a sound accord with the wider world would have us *dance with nature* rather than oppose it.[10] Setting aside the somewhat trite question of not just what nature is but whether it is anything at all,[11] I will submit that *we are the sedentary ones, those who do not dance.* Perhaps we have forgotten how to. Perhaps it is not too late to learn again. In any case, *to go back home, to where you already are,* as Latour enjoins,[12] we—those of us who see that the world is out of joint but are still relatively sheltered from this crisis reality, even if our sense of security is fundamentally false—also face feelings of solastalgia. Only, our pain and pining is not tied to any personal loss, or to any directly encountered injustice, and therefore is not traumatic in any comparable sense. Next to "theirs," our homesickness is vague; it concerns only a felt sense that something is not right; a yearning for a kind of home we have never actually known. How, then, can we get back to where we are without a clear picture of what it is we miss?

---

[8] Ulrich Brand & Markus Wissen, *The Imperial Mode of Living. Everyday Life and the Ecological Crisis of Capitalism* (London: Verso, 2021).

[9] Emil Månsson, "Deadseeing and Its Discontents: A Critical Phenomenology of Ecocide and Earthcare," *Romantik – Journal for the Study of Romanticisms* 10 (2021): 13–34.

[10] Maturana and Verden-Zöller, *The Origin of Humanness in the Biology of Love* (Exeter: Imprint Academic, 2008), 208.

[11] Rather than discard the notion of nature altogether, why not simply agree that a certain historically conditioned conception of it has been and remains inimical to a genuinely ecological thinking? After all, a "love of nature" and the fight for "rights of nature" (with the introduction of ecocide legislation) clearly means something. This is indeed how Husserl dealt with the scientific idea of "material nature" more than a century ago: as a "correlate of the idea of natural science" (in *Ideas Pertaining to a Pure Phenomenology and to a Phenomenological Philosophy*, Book *II* §1, and in *The Crisis of European Sciences and Transcendental Phenomenology* §9). That is, without resorting to the experiential farcicality of claiming there is no nature, when really it is *how we see, think and treat it* that is erroneous and harmful. For a political, historical, and post-colonial account of the Modern view of nature as "brute," see Amitav Ghosh, *The Nutmeg's Curse. Parables for a Planet in Crisis* (John Murray, 2021), e.g., chapter 15: "Brutes," 183–204; or Månsson 2021 (see fn9).

[12] Latour, *After Lockdown*, 52 (see fn5).

Being and Ruin

Where Bruno Latour's Covid-19 book focuses on human metamorphosis, Judith Butler's ditto, titled *What World Is This? A Pandemic Phenomenology,* sets out from the other side of the human-world dance, yet tellingly turns to Max Scheler and especially Maurice Merleau-Ponty, the body phenomenologist, for a meditation on tragedy and grievability. For both Latour and Butler, though, the problem is not the partners but the movement between them; between human bodies that feel, and to a large extent are, out of place, and a world out of joint. It's not really that people and world have lost touch with one another, and if only the two could somehow reconnect… It's rather a sense that we and world are caught up in a rigid, doubly unmoving *dance macabre* of the personal and political, the mental and the material. How might we dispel with this necrotising death dance in favour of truly life-affirming movement? How might we move differently? Butler sets a course for us to follow when they specify that the fact of our interdependency doesn't necessarily equate mutual decency. As they write, "dependency can be miserable; it can take form as exploitation, imprisonment, and legal dispossession; it can be the scene of domination or unwanted self-loss," so that "reciprocity vanishes, as do ideals of equality." Meaning, "the task is less a simple affirmation of interdependency than a collective effort to find or forge the best form of interdependency, the one that most clearly embodies the ideals of radical equality."[13] This is a timely reminder that reconnection isn't guaranteed to bring about blissful coexistence. That it requires concern, care, and responsibility to boot. A few pages before, Butler flags Merleau-Ponty's offhand interpretation of the body-world relation as "one of embrace," for being potentially possessive and smothering. But they fail to see it in, and integrate it into, a more generously embodied context. Butler writes:

> Merleau-Ponty sought to find a language that would challenge the epistemological exercise that presupposes the distance between subject and object. He called intentionality an *embrace* between the subject and world.[14] That was his way of redescribing the correlation between subjective structures of knowing and the countervailing field of the knowable. Consciousness and the world are suited for one another. Consciousness is equipped with structures to know the world as it is, and the world gives itself to consciousness to be known. This adequacy is part of a medieval doctrine of intentionality and survives through the phenomenological legacy. I am not myself convinced by this, but I appreciate the insistence on a harmonious embrace. It is a lovely thought, but what if it turns out to be a chokehold that deals pain rather than joy? What if consciousness and world are singularly ill-suited for one another? Does it not depend on which world, how its limits are established, and how its enduring "facts" are naturalized? Can we even distinguish the world as

[13] Judith Butler, *What World Is This? A Pandemic Phenomenology* (New York: Columbia University Press, 2022), 83.

[14] In *The Visible and the Invisible* (Evanston: Northwestern University Press, 1968), 271. See also Stephen J. Smith's "Gesture, Landscape and Embrace: A Phenomenological Analysis of Elemental Motions," *Indo-Pacific Journal of Phenomenology* 6:1 (2006): 1–10, which Butler also references.

> such from this world, these worlds, striated with inequality, ecological damage, and death-driven capitalism?[15]

To begin at the end, no we cannot distinguish the world as such from everything we are and everything we bring to it and take from it. With this, I'm sure, Merleau-Ponty would agree. It is only fair, then, to remind ourselves that his notion of embrace is taken from a working note scribbled mere months before his untimely death. We can safely *think it further*, therefore, as he surely would have himself, if given the chance. The bond between body and world, between my flesh and the flesh of the world (*chair du monde*), is not only one of embrace, but one of spontaneous embracing and releasing, as is the case, of course, in dance. Sometimes close, sometimes held apart, we move together and, at least if the relation is to be successful, we must react to each other. Butler's worries are poignant and fair, therefore, as a description of a failed relationship to the world—one that is a dominant feature of the imperial mode of living, characterized as it is, using Hartmut Rosa's terminology, by a lack of resonance—but it is hardly fair as a general description of our rapport with the world, neither ideally nor actually, insofar as better bonds require "only" that we move differently and listen carefully. That we practice *attunement*, "a living, dynamic relation with another being," be that another human being, or the earth-ground beneath my feet, or, indeed, a nonhuman other. In any case, "The not-me beckons, making me hesitate. Come."[16] Or, put phenomenologically: affordances are everywhere.

The comfort of the comfort zone, while real in the subjective sense that any emotion is—the proof of solace, after all, is that I am consoled—is nonetheless perceived and, with a technical term, enacted. In short, the comfort zone is maintained by a perception of safety, the basis of which is our everyday enactive rapport with the world.[17] When first we have awoken, that is, we have opened our eyes and minds and hearts to a world imbued with sense; when we see that the world, or reality, is not what we were taught—inert, mute, dead—but is rather pregnant with meaning, in fact personal through and through;[18] when we realize that we co-constitute and co-create Earth's critical zone of life and, moreover, that this zone is haunted by

---

[15] Butler, *What World Is This?*, 78.

[16] Morton, "Attune," 151, 166.

[17] I am thinking here of the discipline of enactivism, which began in embodied cognitive science with Francisco Varela, Evan Thompson and Eleanor Rosch's book, *The Embodied Mind. Cognitive Science and Human Experience* from 1991. For a contemporary overview of enactivism, see Dave Ward et al., "Varieties of Enactivism," *Topoi* 36 (2017): 365–75.

[18] Erazim Kohák explains, "Nor is the overwhelming sense of the clearing as a 'society of persons,' as structured by personal relations, a function of any alleged personality traits of boulders or trees. It is, far more, an acknowledgement of the truth, goodness, and unity of all beings, simply because they are, as they are, each in his [their] own way. That is the fundamental sense of speaking of reality as personal: recognising it as Thou, and our relation to it as profoundly and fundamentally a moral relation, governed by the rule of respect." Kohák, *The Embers and the Stars. A Philosophical Inquiry into the Moral Sense of Nature* (The University of Chicago Press, 1987), 128.

spectral presences;[19] when we see and understand all this, the ground (*sol*) beneath us begins to shift both literally and metaphorically. And our "civilized" understanding of our own being-in-the-world may too. To quote Deborah Bird Rose again: "if we humans are the cause, can we change ourselves enough to change our impacts?"[20] Note that she does not deem humans the problem, only the cause. So, what *is* the problem?

It seems obvious what the problem is: GHG emissions, climate disruption, biodiversity loss, mass extinction. These problems are, however, themselves mere symptoms of the underlying basic problem: that we still haven't figured out how to live at the surface of the earth, with/in the critical zone, and with respect for our planet's material composition, i.e., superficially and materialistically sound, in very specific senses of those words. We have, in short, yet to figure out how to live in this place without ruining it. Without poisoning, demolishing, and exhausting it both at ground and underground level. Without drilling, fracking, extracting, and blowing it up, sucking dry its underland for its black blood, feeding on potash fertilizer for our nutritionally depleted fields, and mining it for the minerals that ignite our in-pocket entertainment systems (iPETS), our tablets and laptops, and other electronic whatnots.[21] And finally, without ceaselessly gushing out toxic gasses into the atmosphere, as parallel to our downward exploits we expand outwards, criss-crossing the skies with planes and drones, launching satellites to live among the stars in order to sustain our terrestrial-imperial mode of living, propped up as it is by screens and celestial information streams that run on the raw materials extracted from below. Which is not merely meant to lament the existence of planes and satellites, or iPhones and laptops. My point is that since we began tearing up the earth—and we have been for a long time, if you can recall the myth of fatefully insatiable prince Erysichthon, the earth-tearer, who ends up eating himself limb for limb—we, which is to say most of us, now have no idea about *how to live at the face of the earth*, which is, should anyone wonder, where it matters most (I refrain from writing "no longer have" from the possible objection that we presumably never knew *perfectly* how to; still, it is safe to say we have never, on the whole, been as clueless as we are today). Today we plainly would not make it without our dead set exhumations of the underland, or our exploits of the ether, even if we wanted to. Or, what's by now considerably worse and far more likely, if we had no choice. Say, if the techno-financial global scaffolding which so much of civilized society banks on came suddenly and violently crashing down as a result of crop failures, resource wars, and natural disasters, or all of it all at once. [*Deep breath. Reset.*]

[19] See, e.g., Jacques Derrida, *Specters of Marx* (New York: Routledge, 2006); Timothy Morton, *Humankind. Solidarity with Nonhuman People* (London: Verso, 2017); Eduardo Kohn, *How Forests Think. Toward an Anthropology beyond the Human* (Berkeley: University of California Press, 2013). Or, for a hands-on example, take the concept of *ghost acreage* in farming.

[20] Deborah Bird Rose, *Wild Dog Dreaming. Love and Extinction* (University of Virginia Press, 2011), 2.

[21] "Atomised, thinking we're engaged when we're pacified/Staring at the screen so we don't have to see the planet die" Kate Tempest, *Let Them Eat Chaos* (Picador, 2016), 67.

Seeing how we enact reality rather than passively finding ourselves thrust into it alters Bird Rose's question slightly, from "Can we change ourselves enough to change our impacts?" to "How can we better manage our interactions with the world and mend our relationship to it?" In Latourian language, this sentiment becomes: "The imperative confronting us, therefore, is to discover *a course of treatment*—but without the illusion that a cure will come quickly," or even that a cure is possible, spurring him on to this ameliorative qualification, that, "If we can't hope to cure ourselves for good, we might at least gamble on the lesser of two evils. After all, one form of treatment entails 'living well with one's ailments,' or even simply 'living well.'"[22] That's why staring down the environmental abyss (cf. Sandberg) and realising our own earthly interbeing as intimately entwined with this planet's myriad other microscopic and macroscopic existences can be equally harmful and reparative. Perhaps one version of this form of treatment could be a *dancing cure* (Kumbaya, I know, but that's another risk you run "preaching" any essentially noneconomistic alternative to standardized GHG mitigation schemes and the like). A dancing cure not only as an act of defiance, and comfort, if one is so inclined, but as an act of practical, secular prayer, not dissimilar to the summoning effects of rain or sun dance. Not just for personal well-being but for communal mending and remoulding, and, why not, for global restoration (*tikkun olam*). Not just literally but as metaphor, as a model even, for those kinds of bonds we need to fashion, or perhaps only unearth and strengthen. Bonds to each other, to ourselves, and to the world. What would they have in common? For one, they would be flexible, forgiving.

This is the qualification I take up here: rather than trying to fix the situation, we must learn or relearn, from with/in our new understanding of the world and ourselves—and whether we favour a Gaian, a bioregional, a so-called kincentric, or some other related conceptualization/cosmology—how to live well in a dying world.[23] It is with this instructive aim in mind, I suggest that dancing as both practice and metaphor might guide us in this endeavour. Help us, as it were, to tread more gently as we skip between and navigate, if not entirely escape, calcified comfort zones of old. And come to live fully, truly and deeply, with/in the dance of relating, to use Donna Haraway's shimmering expression, that we earthlings, human and otherwise, embody and improvise together. "All the actors," Haraway writes, all of us participants in tellurian e-co-existence, "become who they are *in the dance of relating,* not from scratch, not ex nihilo, but full of the patterns of their sometimes-joined, sometimes-separate heritages [embracing and releasing, breath in, breath out…] both before and lateral to *this* encounter. All the dancers are redone through the patterns they enact."[24] Only, the patterns we enact today are getting duller and simpler with every passing moment and every suppressed movement.

---

[22] Bruno Latour, *Facing Gaia. Eight Lectures on the New Climatic Regime* (Cambridge: Polity Press, 2017), 13–14.

[23] This is the title of my PhD thesis (underway): *Living Well in a Dying World. On Humanity's Responsibility for Itself.*

[24] Donna J. Haraway, *When Species Meet* (Minneapolis: University of Minnesota Press, 2008), 25. Notice the *enaction*.

A Dance of Recognition, Reconnection, and Reanimation

But we can change. We are changing. Which brings me at last to Eugene Tovio Gendlin, the backbone thinker of the ECT project to which I have belonged for three years, and his so-called Process Model, introduced in the 1990s. Gendlin writes: "We are learning how to move beyond the old determinism. We are at the beginning of a new science of living process, a more open culture, and a new development of the individual."[25] While the bit about a more open culture is debatable in recent years, it is still true, I think, that significant sections of people are increasingly open and exposed to a new way of thinking about reality, nature, and our place with/in it; about the human body and mind and thinking itself. One such way is seeing how the human body does not end at our skins, a lesson we all had to learn the hard way during the pandemic. That it reaches, as we might say after Bergson and Merleau-Ponty, to the stars.[26] And it's not just that we are made of stardust, as the truism goes—that our bodies are intimately, ontologically linked to elements integrated from our surroundings which are ultimately derived from an interstellar time before biological life (on Earth, at least), and that we depend on the Sun's nuclear furnace and solar wind for our sustenance, and so on—but also, that we serenade them, dream about them, and philosophize with them; that our thoughts and hearts go out to them, and our thoughts and hearts *are* our bodies. As Karel and Iris Schrijver make clear, we are already *living with the stars*.[27] It is only a small step further, then, to commit fully, and *lean into* "dancing with nature," or "dancing with the stars" in the dual sense of the stars as partner and part of myself, moving with/to the *rhythm* of the earth as we settle "out into" the skin of our greater *ecological self*—invoking Freya Mathews' classic work of that name—residing with/in as well as beyond our traditional sense of our selves.[28]

What might a dancing cure prescribe? What ought it accomplish for it to be taken seriously by those of a purely rational mind? The upshot and the prescription in this

---

[25] Eugene Gendlin, "Comment on Thomas Fuchs: The time of the explicating process," in *Body Memory, Metaphor and Movement*, eds. Sabine C. Koch et al. (Amsterdam: John Benjamins Publishing Company, 2012), 74.

[26] Henri Bergson, *The Two Sources of Morality and Religion* (New York: Henry Holt and Company, 1935), 246: "For if our body is the matter to which our consciousness applies itself it is coextensive with our consciousness, it comprises all we perceive, it reaches to the stars." (Quoted a little differently by Merleau-Ponty's editor, in Merleau-Ponty 1968, 57).

[27] Karel Schrijver and Iris Schrijver, *Living with the Stars. How the Human Body is Connected to the Life Cycles of the Earth, the Planets, and the Stars* (Oxford University Press, 2019).

[28] For related aspects of this movement which I cannot begin to explore here, I refer the reader to two more authors also present in this volume. As already mentioned, Timothy Morton explores *attunement* which, along with *resonance*, must accompany any decent dancing ecology. And Michael Marder introduces the concept "the terrene feeling"—a response to Freud's "oceanic feeling"—in "Connectedness," from the Winter 2022 issue of *The Philosopher*. Finally, on *rhythm*, especially in Merleau-Ponty, as it relates to attunement and resonance, see Jessica Wiskus' *The Rhythm of Thought. Art, Literature, and Music after Merleau-Ponty* (The University of Chicago Press, 2013).

case is one and the same: the dancing cure asks us to get moving. To look long and hard, to listen deeply, to tune in, and then to get involved. That is all it wants to accomplish. But that is already a lot. The last thing you need is probably being told to move *more*. Maybe you feel trapped in the proverbial hamster wheel as is, and only wish you could escape or at the very least slow down. That's fair. The dancing cure is not advocating tempo. It's saying… *step back, listen, respond to what is coming to you and coming from you, move with that, from that.* Dancing restores spontaneity. It frustrates convention. It kicks trite custom to the curb. Or so I conjecture.

As I have no practical experience to draw on, a word on how my ignorance and ineptitude in the dance department might be spun as a positive seems in order. I represent the imperial, sedentary mode of living as well as anyone. A Caucasian, Scandinavian, cissexual, able-bodied, well-educated man, I belong to the Majority Gold Club of this world. I ooze privilege. And while I do believe that "we" should first and foremost listen to and learn from the world's five percent Indigenous peoples, who "somehow" manage not to destroy their environments as a way of life, it is no less important to acknowledge that it is we/I who are in the wrong who must do the work and learn the steps. It is we who must move differently with/in/through this world, *I* who must change *my* ways. As bell hooks' 2004 book on men, masculinity, and love puts it, it is we who must mobilize *the will to change*.[29] The first step is showing up, mind open and body ready. And who better to train than a complete novice, asking: How do I learn to move differently through this dying world, when all I'm taught and asked to do is use and abuse? If I can, anyone can. I'm not saying I have learned, only that I am as good—read, as corrupted—a candidate for a complete turnaround as anyone. In short, this is a self-help text as much as anything: me helping my own understanding in order for me to move differently.

So, what have I learned? That I don't need to come out so much as realize that I am already out. That safe space is an illusion, and a perfect body too. Earthly existence is awkward. Embracing the awkward is how you let go of it.[30] Dance even if it doesn't come naturally. Don't mind the body-mind gap; it's overstated anyways, and falling through it will only set you free. Dance because it is childish. Play with it. Dance because there's nothing more serious or vitalising. Dance to know. Dance to connect. Dance to heal.[31]

All the dancers are redone through the patterns they enact, writes Haraway. We are all dancers, whether we know it or not, and whether we like it or not. "I don't

---

[29] bell hooks, *The Will to Change* (New York: Atria Books, 2004).

[30] Because attentive and spontaneous movement is the antithesis to embarrassment: from Portuguese *embaraçar*, from *baraço*, 'halter', the rope used to tether a horse. Only movement can *re-lax* my ingrained restraints. Learning by doing.

[31] See Eugen Fink, *Play as Symbol of the World. And Other Writings* (Bloomington: Indiana University Press, 2016); Paul Shepard, *Nature and Madness* (Athens: University of Georgia Press, 1998), specifically chapter six: "The dance of Neoteny and Ontogeny"; and, with moving acknowledgement, Kimerer L. LaMothe, *Why We Dance. A Philosophy of Bodily Becoming* (New York: Columbia University Press, 2015).

dance" just won't cut it. The only choice we have now is how to move. Awkwardly: rigidly, contrarily, barely; standing still but trembling, the world rendered opponent.[32] Or evocatively: tenderly, attentively, the world *held* as a genuine partner. Albrecht writes that, "The word 'emotion' has its origins in the Latin *movēre*, 'to move,' and *ēmovēre,* 'to agitate, disturb,'" and this, he asserts, "makes the universe an 'emotional' place."[33] Dancing is a, if not *the* way the human animal formalizes and channels this universal trend. Dancing does not simply portray or exhibit human emotion. Dancing *is* emotioning, as Maturana and Verden-Zöller might say. Or dancing *is* living process, with Gendlin. In Maturana and Verden-Zöller's words, "a change of emotion or mood is a change of brain and body. Through different emotions human and non-human animals become different beings, beings that see differently, hear differently, move and act differently."[34] In keeping with Bergson's astral metaphor, we might say that a person's *atmosphere* changes—we say this already of a social situation or of physical space—as their mood changes. But just like a change of emotion and mood is also a change of brain and body, we know the influence works in the other direction too; that we can affect an emotional change simply by moving and acting differently. Motion can move us emotionally. This is why we go for walks when we are down, or run, or dance, or swim, or whatever it is we do. This is, effectively, us dancing with/in/from our greater selves—what Gendlin calls en#2, the world that is constituted by the body-environment interactions of all Earthlings who somehow partake in "my" body-environment, from microbiota to crisp morning air to Cartesian mind. This is a dance of brain and behavior, to use those old terms, which is part of the greater choreography to which we all belong, implying moods and emotions as well as gestures, postures, prospects, and whatever else pertain to "us" at any one moment. As Maturana and Francisco Varela wrote already in 1987:

> Whatever we do in every domain, whether concrete (walking) or abstract (philosophical reflection), involves us totally in the body, for it takes place through our structural dynamics and through our structural interactions. Everything we do is a structural dance in the choreography of coexistence. […] a social dynamics which points up a basic ontological feature of our human condition that is no longer a mere assumption, that is, *we have only the world we bring forth with others, and only love helps us bring it forth.*[35]

This is love, mind you, of a very unusual, rather technical kind. To Maturana, romantic love, parental love, and filial love; love of country, love of God, love of nature, and all the rest; all of those are not so much different forms, kinds, or levels of love, but "in fact connote only different relational dimensions of our living as loving animals." With his go-to expression of what love is, he sums up his views

---

[32] See, for example, Hartmut Rosa, *The Uncontrollability of the World* (Cambridge: Polity Press, 2020): Chapter 1: The World as a Point of Aggression, 5.

[33] Glenn A. Albrecht, *Earth Emotions. New Words for a New World* (Ithaca: Cornell University Press), 1.

[34] Maturana and Verden-Zöller, *The Origin of Humanness…*, 38.

[35] Humberto R. Maturana and Francisco J. Varela, *The Tree of Knowledge. The Biological Roots of Human Understanding*, revised edition (Boulder, Colorado: Shambhala Publications, 1998), 248.

thusly: "Love is the domain of those relational behaviors through which the other arises as a legitimate other in coexistence with oneself." [36] Love it seems, being a *domain of relational behaviors*, is another word for attunement and resonance. For proper recognition. For truly seeing another, taking them for and letting them be who they are. Which in turn lets *me* be who *I* am. Our project in Embodied Critical Thinking has foregrounded various methods for bringing out pre-reflective knowledge and engaging one's felt sense precisely because they end up altering not just the-thought-about but the thinker. They alter *how* we think and how we *want* to think. A dancing cure is, in short, the best way I can think of to counter the default mode of Deadseeing with/in your-larger-self. Donata Schoeller and Neil Dunaetz summarize the process well with regards to Gendlin's thinking-practice:

> Eugene Gendlin's philosophical project [...] teaches us that experience cannot be understood as only an object of theories. He demonstrates instead how to systematically think not only *about*, but *in* and *with* experience. The radically reflective style of Gendlin's thinking shows how next moves and precise symbolic distinctions *emerge*, affecting the process they emerge from. Gendlin understands this process as instancing an order he calls "responsive." Understanding its implicit precision needs one's engaging it. [And from a few paragraphs later:] Practices involving reflective interaction with what is experientially implicit have the quality of dance moves: how one moves in relation to, is formed by, and is a change in, what one moves with.[37]

Human, Animal, Elemental

Let us rehearse the main ways of relating: to surroundings, to others, to oneself. Here, then, is a non-exhaustive list of levels of relating in which to move more or less *dancingly*, which is to say, lovingly:

- Dancing literally: with a single partner or with several; romantically or otherwise.
- Dancing with a partner metaphorically, say, a spouse or a friend; in life, through life.
- Dancing with/in one's lifeworld; balancing family, work, hobbies, chores, dreams etc.
- Dancing with oneself; moving back and forth with oneself, vertically in any one moment, and horizontally through life; constantly steadying and checking one's thoughts and body.
- Dancing in relation to the earth-ground that carries me; paying more or less attention to it, being more or less attuned to it; responding to it or not, knowing, or not, that I am, or not.

[36] Maturana and Verden-Zöller, *The Origin of Humanness*..., 223.

[37] Donata Schoeller and Neil Dunaetz, "Thinking emergence as interaffecting: approaching and contextualizing Eugene Gendlin's Process Model," *Continental Philosophy Review* 51 (2018): 124.

- Dancing with nature in the sense of feeling into (empathy) and responding (giving back) to what moves all around me and through me; my ecological self, the flesh of the world.
- Dancing with nature with others; being part of a collective we, finding common ground from/on which to dance together; with/for each other, with/for *this* place, with/for nature.

Dancing is global. If we allow it—meaning if we acknowledge it *and* sustain it—everything dances. Which is to say, everything has the potential to. Consider, with David Abram, the migratory magic of sandhill cranes, wild salmon, and monarch butterflies. As these feral beings "are riding waves of sensation, responding attentively to allurements and gestures in the topological manifold, reverberating subtle expressions that reach them from afar," they are in effect "dancing not with themselves but with the animate rondure of the Earth, their wider flesh!"[38] Or take Karen Barad's persistent use of a dancing language vis-à-vis quantum physics, as in, "entanglements in the lively dance of mattering" and "the ongoing dance of agency immanent in [the manifold's] material configuration."[39] *Reality dances*, strung out between macro and micro levels of the *elemental* both "in the sense it was used to speak of water, air, earth, and fire, that is, in the sense of a *general thing*, midway between the spatio-temporal individual and the idea, a sort of incarnate principle that brings a style of being wherever there is a fragment of being";[40] and in the chemical sense of a primary constituent of matter. In both instances, the question remains: how are we justified in using dancing quite so… poetically?

Seeing that the world aka reality aka nature is not radically apart from us or wholly outside of us implies a metaphysical belief. It tells us that nonhuman reality is not just a passive backdrop to our active lives. That it speaks to us. In the words of Hans Jonas, "*Being, in the testimony it gives of itself, informs us not only about what it is but also about what we owe to it.*"[41] Such metaphysical bearings are required for the natural philosopher to elevate her thinking from poetic opining to serious analysis. But more than that, it tells us how "things" relate, which is what ontology does. Akin to Jonas' credo is James J. Gibson's definition of what he terms *affordance*. He writes: "Perhaps the composition and layout of surfaces *constitute* what they afford. If so, to perceive them is to perceive what they afford. This is a radical hypothesis, for it implies that the 'values' and 'meanings' of things

[38] David Abram, "Wild ethics and participatory science: thinking between the body and the breathing earth" in *Kinship. Belonging in a World of Relations, Volume 1: Planet* (Libertyville, IL: Center for Humans and Nature Press, 2021), 60.

[39] Karen Barad, *Meeting the Universe Halfway. Quantum Physics and the Entanglement of Matter and Meaning* (Duke University Press, 2007), 37, 246. For a popular account of dance cosmology—from "primitive" to modern physics—see Marcelo Gleiser, *The Dancing Universe. From Creation Myths to the Big Bang* (Dartmouth College Press, 2005).

[40] Merleau-Ponty, *The Visible and the Invisible*, 139–40.

[41] Hans Jonas, "Towards an Ontological Grounding of an Ethics for the Future," in *Mortality and Morality. A Search for the Good after Auschwitz* (Evanston: Northwestern University Press, 1996), 101. Italicized in the original.

[including animal, vegetal, and possibly spectral "things" I would claim] in the environment can be directly perceived."[42] This idea is radical, no doubt, in the Western canon. But not so within non-Western thought systems. It is obvious to most, I take it, that dancing played a pivotal role in so-called primitive human cultures, and indeed still does to cultures that have preserved some remnants of our pre-civic ways. We talk of folk dance and file it alongside folklore, which is just a pace away from myth or outright fantasy. Or so we believe. But what if dancing speaks to something profoundly human, as no lesser a hero of the Western canon than star-crazed philosopher of dance, Friedrich Nietzsche, believed? What if dancing is the most human, nay, the most more-than-human *creaturely* language that exists? What if it is both, what if it is all: human, animal, elemental, Earth-communication *par excellence*? Communion-ising, sol-idarity. In *Human, All Too Human*, Nietzsche speaks of dance as a language of gesture, older than its spoken or written counterparts, and relates it to our involuntary propensity for mimicry of gestures. (What is this if not a nineteenth century version of the 1990s cognitive science popular phenomenon mirror neurons?). This dance is precisely what Gendlin coveys under the heading The Dance, on the behavioural back-and-forth that ensues from a meeting between two monkeys, when instead of fighting or fleeing they simply… relate. What happens is that one threat gesture will "alter [the other] monkey's bodylook, sound, and movement somewhat," and that "will carry the first monkey forward somewhat differently, and so on." Which leads to "a sequence of slightly different *versions* of the same 'gesture.' I call this *a dance*."[43] A dance of primates, a category we ourselves of course belong to.

From singular encounter, Haraway extrapolates to structural level: "The partners do not precede the meeting; species of all kinds, living and not, are consequent on a subject- and object-shaping dance of encounters."[44] The same goes for the human animal. We too rely on this dance of encounters. It is even in our DNA. From Gendlin's commentary on Thomas Fuchs' example of learning to waltz:

> Certainly [waltzing] is a *learned* rhythm, but we could not learn it if the body's living process didn't already have some rhythms or the capacity for rhythm. Some person's body process *first generated* the waltz pattern. Then it was divided into parts, so that now others can learn part-moves until the rhythm emerges for them too. It emerges as a new form reshaping the human body's rhythms. And because it can emerge from human bodies, therefore everyone *can* learn it. [Thus:] *Dancing is not just the learned steps. Dancing is the original body process now further developed.*[45]

Dancing, it seems, is primaeval. That doesn't mean outdated, though, for if it is not *further developed*, literally and metaphorically, we stand still; which in evolutionary

[42] James J. Gibson. *The Ecological Approach to Visual Perception* (Boston: Houghton Mifflin, 1979), 127. See also, Artese and Kiverstein, "The Social Orders of Existence of Affordances," *Philosophia Scientiæ* 26:3 (2022/3): 211–32.

[43] Gendlin, *A Process Model* (Northwestern University Press, 2018): Chap. VII, subsection "(b) The Dance," 116–117.

[44] Donna J. Haraway, *When Species Meet* (Minneapolis: University of Minnesota Press, 2008), 4.

[45] Gendlin, "Comment on Fuchs," 75.

terms amounts to regress. And not just that, but "we learn reading and writing like we learn the waltz," Gendlin continues. From which it is not, perhaps, outlandish to think that what is true of dancing, reading, and writing, is also true of thinking itself? In the words of Nietzsche again, "thinking wants to be learned as dancing wants to be learned, *as* a kind of dancing…"[46] If only we could help move it along, dancingly. That has been the aim of our ECT project.

Let me by way of conclusion reiterate: the beauty of the new kind of thinking that we practice in Embodied Critical Thinking is, for me at least, that it never ends. That it "needs one's engaging it," even when it is awkward and difficult. Like phenomenology before, like dancing itself, it is creatively open, flowing and merging, in living process, *here* but not *just* here. It's an ecological thinking that considers the intimately personal in order to reach the most universal and common, realising that how we move in the world—how we comport ourselves, including how we think and perceive reality and ourselves—is ultimately what matters most in this world. Not only for ourselves but *for the world.*

Bibliography

Abram, David. 2021. Wild ethics and participatory science: thinking between the body and the breathing earth. In *Kinship. Belonging in a World of Relations, Volume 1: Planet*, ed. Gavin van Horn, Robin Wall Kimmerer, and John Hausdoerffer, 50–62. Libertyville: Center for Humans and Nature Press.

Albrecht, Glenn A. *Earth Emotions. New Words for a New World*. Ithaca: Cornell University Press.

Barad, Karen. 2007. *Meeting the Universe Halfway. Quantum Physics and the Entanglement of Matter and Meaning*. Duke University Press.

Bergson, Henri. 1935. *The Two Sources of Morality and Religion*. New York: Henry Holt and Company.

Brand, Ulrich, and Markus Wissen. 2021. *The Imperial Mode of Living. Everyday Life and the Ecological Crisis of Capitalism*. London: Verso.

Brown, Charles S., and Ted Toadvine, eds. 2003. *Eco-Phenomenology. Back to the Earth Itself.* Albany: SUNY Press.

Butler, Judith. 2022. *What World Is This? A Pandemic Phenomenology*. New York: Columbia University Press.

Derrida, Jacques. 2006. *Specters of Marx*. New York: Routledge.

Fink, Eugen. 2016. *Play as Symbol of the World. And Other Writings*. Bloomington: Indiana University Press.

Garland-Thomson, Rosemarie. 2020. Misfitting. In *50 Concepts for a Critical Phenomenology*, ed. Gail Weiss, Ann V. Murphy, and Gayle Salamon, 225–230. Evanston: Northwestern University Press.

Gendlin, Eugene. 2018. *A Process Model*. Evanston: Northwestern University Press.

———. 2012. Comment on Thomas Fuchs: The time of the explicating process. In *Body Memory, Metaphor and Movement*, ed. Sabine C. Koch, Thomas Fuchs, Michela Summa, and Cornelia Müller, 73–81. Amsterdam: John Benjamins Publishing Company.

Ghosh, Amitav. 2021. *The Nutmeg's Curse. Parables for a Planet in Crisis*. John Murray.

[46] Friedrich Nietzsche, *Twilight of the Idols. Or, How to Philosophize with the Hammer* (Indianapolis: Hackett Publishing Company, Inc., 1997 [1889]), 48. Translation by Richard Polt.

Gibson, James J. 2014 [1979]. *The Ecological Approach to Visual Perception*. Boston: Houghton Mifflin.
Gleiser, Marcelo. 2005. *The Dancing Universe. From Creation Myths to the Big Bang*. Hanover: Dartmouth College Press.
Haraway, Donna J. 2008. *When Species Meet*. Minneapolis: University of Minnesota Press.
hooks, bell. 2004. *The Will to Change*. New York: Atria Books.
Husserl, Edmund. 1989. *Ideas Pertaining to a Pure Phenomenology and to a Phenomenological Philosophy. Second Book: Studies in the Phenomenology of Constitution*. Trans. Richard Rojcewicz and André Schuwer. Dordrecht: Kluwer Academic Publishers.
———. 1970. *The Crisis of European Sciences and Transcendental Phenomenology*, An Introduction to Phenomenological Philosophy. Trans. David Carr. Evanston: Northwestern University Press.
Jonas, Hans. 1996. Toward an Ontological Grounding of an Ethics for the Future. In *Mortality and Morality. A Search for the Good after Auschwitz*, ed. Lawrence Vogel, 99–112. Evanston: Northwestern University Press.
Kohák, Erazim. 1987. *The Embers and the Stars. A Philosophical Inquiry into the Moral Sense of Nature*. The University of Chicago Press.
Kohn, Eduardo. 2013. *How Forests Think. Toward an Anthropology beyond the Human*. Berkeley: University of California Press.
LaMothe, Kimerer L. 2015. *Why We Dance. A Philosophy of Bodily Becoming*. New York: Columbia University Press.
Latour, Bruno. 1993. *We Have Never Been Modern*. Cambridge, MA: Harvard University Press.
———. 2017. *Facing Gaia. Eight Lectures on the New Climatic Regime*. Cambridge: Polity Press.
———. 2018. *Down to Earth. Politics in the New Climatic Regime*. Cambridge: Polity Press.
———. 2021. *After Lockdown*. Cambridge: Polity Press.
Latour, Bruno, and Peter Weibel, eds. 2020. *Critical Zones. The Science and Politics of Landing on Earth*. Cambridge: MIT Press.
Leopold, Aldo. 1987 [1949]. *A Sand County Almanac. And Sketches Here and There*. Oxford University Press.
Maclaren, Kym. 2014. Intimacy and embodiment: An introduction. *Emotion, Space and Society* 13: 55–65.
———. 2018. Intimacy as Transgression and the Problem of Freedom. *Puncta: Journal for Critical Phenomenology* 1: 18–40.
Marder, Michael. Connectedness. *The Philosopher* (Winter 2022 issue: *The New Basics: Planet*).
Maturana, Humberto R., and Francisco J. Varela. 1998. *The Tree of Knowledge. The Biological Roots of Human Understanding*, revised edition. Boulder: Shambhala Publications.
Maturana, Humberto R., and Gerda Verden-Zöller. 2008. *The Origin of Humanness in the Biology of Love*. Exeter: Imprint Academic.
McRuer, Robert. 2020. Compulsory Able-Bodiedness. In *50 Concepts for a Critical Phenomenology*, ed. Gail Weiss, Ann V. Murphy, and Gayle Salamon, 61–68. Evanston: Northwestern University Press.
Merleau-Ponty, Maurice. 1968. *The Visible and the Invisible*. Evanston: Northwestern University Press.
Morton, Timothy. 2017a. *Humankind. Solidarity with Nonhuman People*. London: Verso.
———. 2017b. Attune. In *Veer Ecology. A Companion for Environmental Thinking*, ed. Jeffrey Jerome Cohen and Lowell Duckert, 151–167. Minneapolis: University of Minnesota Press.
Månsson, Emil. 2021. Deadseeing and Its Discontents: A Critical Phenomenology of Ecocide and Earthcare. *Romantik – Journal for the Study of Romanticisms* 10: 13–34.
National Research Council. 2001. *The. Basic Research Opportunities in Earth Science*. Washington, DC: The National Academies Press.
Nietzsche, Friedrich. 1997 [1889]. *Twilight of the Idols. Or, How to Philosophize with the Hammer*. Trans. Richard Polt Indianapolis: Hackett Publishing Company, Inc.
Rosa, Hartmut. 2020. *The Uncontrollability of the World*. Cambridge: Polity Press.

Rose, Deborah Bird. 1996. *Nourishing Terrains. Australian Aboriginal Views of Landscape and Wilderness*. Canberra: Australian Heritage Commission.

———. 2011. *Wild Dog Dreaming. Love and Extinction*. University of Virginia Press.

Schoeller, Donata, and Neil Dunaetz. 2018. Thinking emergence as interaffecting: approaching and contextualizing Eugene Gendlin's Process Model. *Continental Philosophy Review* 51: 123–140.

Schrijver, Karel, and Iris Schrijver. 2019. *Living With the Stars. How the Human Body is Connected to the Life Cycles of the Earth, the Planets, and the Stars*. Oxford University Press.

Shepard, Paul. 1998. *Nature and Madness*. Athens: University of Georgia Press.

Smith, Stephen J. 2006. Gesture, Landscape and Embrace: A Phenomenological Analysis of Elemental Motions. *Indo-Pacific Journal of Phenomenology* 6 (1): 1–10.

Tempest, Kate. 2016. *Let Them Eat Chaos*. Picador.

Varela, Francisco J., Evan Thompson, and Eleanor Rosch. 2016. *The Embodied Mind. Cognitive Science and Human Experience*, revised edition. Cambridge, MA: MIT Press. [1991].

Ward, Dave, David Silverman, and Mario Villalobos. 2017. Introduction: Varieties of Enactivism. *Topoi* 36: 365–375.

Wiskus, Jessica. 2013. *The Rhythm of Thought. Art, Literature, and Music after Merleau-Ponty*. The University of Chicago Press.

Emil Månsson is a doctoral candidate in the interdisciplinary project on Embodied Critical Thinking based out of University of Iceland's department of philosophy. His work is premised on the equally heartbreaking and incriminating reality of our planetary murder. What ideas and beliefs underwrite it, and can we move against it? Having spent three metamorphozing years on a raw mid-Atlantic island, marked by the Covid-19 pandemic and a rare 6-months long volcanic eruption near the capital of Reykjavík, Emil now resides in the Danish intentional community of Svanholm with his wife and children, where he tries and frequently fails to embody a truly attuned terpsichorean theory-practice.

Chapter 12
The Embodied, Experiential and Elemental Turn in Environmental Ethics

Sigridur Thorgeirsdottir

Abstract In this paper I reflect my experience of teaching "Ethics of Nature" or environmental ethics for over 15 years. Environmental ethics is the most groundbreaking branch of contemporary ethics because of its environmentally embedded and embodied conception of human beings. For a change towards a more sustainable culture to take place it is however not enough that our understanding of our place in the environment becomes more ecocentric and to realize how human life interconnected with all life. It is also not sufficient that the big economic, financial systems and production systems that constitute an unsustainable economy become less unsustainable. A change in the way we experience and sense ourselves as environmental beings also needs to happen. I respond to voices within environmental ethics that call for anthropocenean"sensibilities" like Nancy Tuana does and Donna Haraway does with her idea of "staying with the trouble." Deborah Bird Rose also calls for a deeper sense of listening to the non-human environment. For the cultivation of practices of such sensibilities and listening I have been inspired by methodologies of embodied and enviromental thinking developed within enactivist, processual phenomenological and pragmatist philosophies, especially by Eugene Gendlin and by Claire Petitmengin. Such methodologies of embodied thinking and understanding have hence been integrated in my teaching of environmental ethics and I view them as contributing to the "Inner Development Goals" that are seen as complimentary to the UN Sustainability Development Goals.

Keywords Embodied thinking and understanding · Methodologies of embodied thinking and understanding · Paradoxes of living in times of climate change

Environmental ethics is the most radical and groundbreaking branch of contemporary ethics. With its relational, embedded, and embodied idea of the human being, high hopes have been tied to it. Freya Matthews precisely hoped that an ecocentric

S. Thorgeirsdottir (✉)
Department of Philosophy, University of Iceland, Reykjavik, Iceland
e-mail: sigrthor@hi.is

L. Škof et al. (eds.), *Elemental-Embodied Thinking for a New Era*, Sophia Studies in Cross-cultural Philosophy of Traditions and Cultures 42,
https://doi.org/10.1007/978-3-031-42119-8_12

outlook as an inspiring way of thinking about human beings place on this earth among other living beings would change the way humans treat the natural environment. However, environmental ethics has not lead to a widespread reorientation of values. Neither did greater familiarity with indigenous ways of environmental thinking lead to extensive establishment of ecological communities, making Matthews ask whether environmental ethics has failed.[1] In order for a change in culture and mentality to take place it is not enough that our understanding of our place in the environment becomes more ecocentric in the sense of understanding how human life interconnected with all life. It is also not enough that the big economic, financial systems and production systems that constitute an unsustainable economy change. A change in the way we experience and sense ourselves as environmental beings also needs to happen. This is my experience after having taught a course on "Ethics of Nature" at my university for over 15 years.[2] Here I therefore respond to voices within environmental ethics that call for anthropocenean "sensibilities" like Nancy Tuana does and Donna Haraway does with her idea of "staying with the trouble." For the cultivation of such sensibilities I have been inspired by methodologies of embodied and enviromental thinking developed within recent phenomenological and pragmatist philosophies, especially by Eugene Gendlin and by Claire Petitmengin.[3] I view these methodologies as contributing to the demand for "Inner Development Goals" that are seen as complimentary to the UN Sustainability Development Goals (https://innerdevelopmentgoals.org). I however want to begin to discuss how both some of the main theoretical developments in the field of environmental ethics of the last decades and common everyday experiences of the environemental predicament have lead to the elemental and embodied turn within environmental ethics that I present here.

---

[1]Freya Matthews, "We've had forty years of environmental ethics – and the world is getting worse," https://www.abc.net.au/religion/weve-had-forty-years-of-environmental-ethics%2D%2D-and-the-world-is-/10214334.

[2]I would like to express my gratitude to the forerunners and co-teachers who contributed to the development of this course as well as to my esteemed colleagues working on philosophy of nature at the University of Iceland, Þorvarður Árnason, Guðbjörg R. Jóhannsdóttir, Ólafur Páll Jónsson, Jón Kalmansson, William Konchak, Emil Månsson, Ole Sandberg, the late Páll Skúlason, Björn Þorsteinsson. Additionally, I am thankful to numerous students of the course on Ethics of Nature, (many of them travelling from abroad to study environmental issues in Iceland), who inspired me to contemplate environmental ethics through the lens of Icelandic geology and landscapes. Lastly, I have drawn inspiration from my companions in the environmental movement who fight the environmental destruction happening in Iceland wrought by corporate greed and political shortsightedness.

[3]See Nancy Tuana, "From a Lifeboat Ethic to Anthropocenean Sensibilities," *Environmental Philosophy* 17, no. 1 (2020): 101–123; Donna Haraway 2016, *Staying with the Trouble: Making Kin with the Chthulucene* (Durham: Duke University Press); https://innerdevelopmentgoals.org

The Experiential and Embodied Turn

Students of Ethics of Nature need to be met where they experientially sense environmental challenges in their own lives and behavior. There is an elephant in the classroom of conventional approaches within environmental ethics where students learn *about* theories and different ways of applying ethical theories by reflecting concrete environmental debates and controversies. That elephant is an essential starting point of an ethics of environmental issues precisely because it is often not consciously perceived or not explicitly attended to since it is like a low voice in the back of one's head that one is not attuned to listening to. Attending to this quiet, elusive sense involves confronting oneself in the face of environmental challenges and experiencing it as being torn. This disposition captures the experience of living and experiencing the contradictions inherent in responding to environmental challenges. For most people this is like living a paradox. Firstly, one may have knowledge about the challenges, but one's practices feel inadequate and lagging behind. The problem of climate change is massive, and despite the scientific knowledge and frequent news of climate-related disasters, people often don't directly feel the impact of climate change, making it difficult to fully comprehend and grasp its severity. "Hyperobjects," the term that Timothy Morton coined for enormous phenomena such as the climate, plastic in the ocean, and pollution, are so big and unfathomable, making individuals feel like their actions for the environment matter so little and are mostly done for the sake of their own conscience. [4] The problem feels of the magnitude of having to move a mountain. The crux can also be felt more desperately, leaving some with a climate anxiety that in extreme cases leads to a flight or freeze behavior. Common to extreme reactions is a disconnection, be it feeling paralyzed or fleeing the issue at hand. It is a disconnect to the sentiment that evokes the different reactions making one unable to process it. For this reason, the teaching, the readings, the assignments of our course on Ethics of Nature have taken an embodied, experiential, and elemental turn. More importantly, overly rationalistic debates come to feel like part of the problem because talking "about" an issue in abstract terms of duties, rights, virtues, benefits, principles, inequalities, etc. feels as if one is analyzing an issue that is external to oneself, like an object under an enlarging glass.

There are radical reactions to climate change challenges that are for many too extreme to follow. With his book, *How to Blow up a pipeline: Learning to Fight in a World on Fire*, Andreas Malm (2021) represents the view that sabotage and damage of property are justified and needed. With his paper on "deep adaption," Jem Bendell has launched a social movement of people who expect a societal collapse due to effects of climate change and have begun to prepare themselves to adapt to it,

[4] Timothy Morton, *Hyperobjects: Philosophy and Ecology after the End of the World* (Minneapolis: University of Minnesota Press, 2013).

guided by values of nonviolence and compassion.[5] The approach adopted with embodied thinking consists in beginning by acknowledging the discomfort living in the crux of knowing a problem and yet to feel quite powerless to do something effective about it is a point of departure for that is indeed a much more common disposition than taking action in a radical manner.

Discomfort

One of the aspects of discomfort is the very fact that one can live quite comfortably in times of climate change with its predictions of changes on a massive scale, increasing frequency of natural disasters, and even impending social collapse. Yet, living comfortably is not like living in climate change denial but somehow an in-between state between rejecting that it is happening and climate anxiety because the effects of climate changes. One goes about one's life, travelling, driving, consuming, working, knowing that most of the things one does are part of an economy that is unsustainable, consuming products that are grown on earthly soil that is "exhausted" due to overproduction, depending on water reserves that diminish faster than they can be replenished, a monetary system that is out of touch with its earthly limitations, a debt-economy spiraling upwards increasing the gap between the very few hyper-wealthy and the rest, a means of production of goods and food that is economically, politically, and healthwise unviable, unjust, and repressive and to the plight of our fellow species that share the planet with us.

Concepts of "just," "fair," "equal," "care," etc. begin to sound like low intensity tones against the forceful sound of the base and the beat of the drums of the powers that be that make our massive, globally interconnected supply and service systems move, economically highly functionally but ecologically absolutely disfunctionally. These ethical concepts however sound better when it is kept in mind that indicators of quality of life such as life expectancy, literacy rates and the alike are for the most part better than they have ever been. A continued upward tendency of these indicators is however now severely threatened by the effects of climate change that are becoming more and more apparent and more frequently experienced. Like an old person once said to me with a slightly guilty sounding tone in the voice, "I am relieved that I will not live long enough to have to face the environmental challenges younger generations have to deal and come to terms with."

Even if a person does not believe that climate change is to a large extent man made, the effects of higher temperatures, the fact that 25,000 species go extinct every year, the growing number of climate refugees, and the need for ways and means to deal with the effects of these developments are undeniable. So, one not only lives in the crux of discomfort of feeling impotent against overbearing

---

[5] Jem Bendell, *Deep Adaption: A Map for Navigating Climate Tragedy*, 2020. Available in different versions here: https://lifeworth.com/deepadaptation.pdf.

systems but also dies in it, often with a regret of not having done enough to create less environmentally detrimental conditions for one's offspring, and with a feeling of having left a mess for the following generation. Whether one blames "climate deniers" or "climate industry" that is accused of making a profit to the benefit of some financial powers, one has to "stay with the trouble" and sense into "how it is" to be environmentally and ecologically attuned. Just as everybody, else even climate deniers have become much more aware of our environmental dependency and that marks the embodied, experiential and elemental turn in the teaching of Ethics of Nature.

Empathic Understanding

What is common to theories of environmental ethics, be they anthropocentric, eco-centric, or indigenous theories, is that they require one to think of and imagine oneself as part of nature and the environment and standing in relation with other non-human living beings. Ethical theories about animals necessarily have an anthropocentric point of departure, requiring humans to extend their moral circle to non-human animals by granting them rights.[6] Peter Singer bases his idea on the need to grasp and understand the preferences of non-human animals, what they want and desire as sentient beings for their living.[7] With her recent plead for justice for animals, Martha Nussbaum enriches her justice based ethical perspective by calling for a more compassionate approach towards animals that turns human beings from exploiters and users to true friends of animals.

With the recent growth of theories of vegetal being and plant thinking one is encouraged to learn from how plants "think" (although the word "think" may not apply to plants when compared to human thinking).[8] This approach implies that one tries to understand better how plants function, struggle, and thrive by way of learning more about findings of botany and biology. One is also encouraged to make an imaginative effort, to so to speak "put one self in the shoes" of a non-human, vegetal other. Such an effort implies cognitively grasping processes of how plants communicate with one another and how they interact with their surroundings in ways that enable them to hold their ground, defend themselves, survive, accommodate and support other plants in their growth. Yet, a human being can neither "think" nor "experience" like a plant. On a deeper level it is perhaps possible to bodily sense

[6] Tom Regan, *Defending Animal Rights* (Champaign, IL: University of Illinois Press, 2000).

[7] Peter Singer, *Animal Liberation: A New Ethics for our Treatment of Animals* (New York: Random House, 1975).

[8] Luce Irigaray and Michael Marder, *Through Vegetal Being: Two Philosophical Perspectives* (New York: Columbia University Press, 2016); Michael Marder, *Plant-Thinking. A Philosophy of Vegetal Life* (New York: Columbia University Press, 2013).

vegetal levels of one's being with interoceptive awareness of inner body vegetal processes or by meditating deeply on plant life out there. It is nevertheless impossible to transcend a species barrier experientially, although one can understand preferences of sentient non-human being. Even a cognitively compassionate attitude like Nussbaum calls for is rooted in an ethically empathic understanding that allows us to have insight into preferences of non-human animals and be amazed by their way of living, their different and ingenious forms of intelligence, awakening moral compassion for our companion species.

The distinction between cognitive and moral empathic understanding that dates back to Edith Stein's phenomenology of empathic understanding is helpful to understand the approach that Nussbaum adopts to combine justice and compassion for non-human animals.[9] Nussbaum's idea of compassion is rather on the moral side of empathy which means that an empathic understanding of animals motivates to act morally. Her concept of compassion has a cognitive side, as the ability to grasp and understand another's consciousness on the basis of an ability to sense into it (*Einfühlung*) without necessarily being morally motivated to do something, and obviously human ability of "Einfühlung" with non-human beings is limited by the species barrier.

A second distinction between sensual and emotional empathy can further help to hold apart different degrees of empathic understanding of non-human others and it also delineates different steps of attuning to non-human nature.[10] The embodied and experiential turn within environmental ethics introduced here extends the scope of cognitive understanding by arguing for the necessity of a more embodied cognitive understanding that is empathic, sensual and affective. Such an understanding can connect us more deeply with non-human life, and also with sensual, animalic, vegetal and even mineral levels of living processes in ourselves both for purposes of having more affinity with other living beings and for sensing more intensely our natural belonging as a process of living. Moral empathy therefore needs to be grasped and understood in its relation to cognitive empathy.

Aesthetic and Calculative Approaches

A stepping stone towards introducing theories and practices of embodied thinking and understanding as a methodology of the course on Ethics of Nature consisted in introducing aesthetics of nature. Students did exercises in experiencing landscapes and man made environments as well as reflecting on environmental art. For example, they did an exercise that consisted in walking through a natural landscape and

[9] Edith Stein, "On the Problem of Empathy," in *The Collected Works of Edith Stein, Vol. 3*, translated by Waltraut Stein (Washingtond DC: ICS Publications, 1989).

[10] Fredrik Sveneaus, "Edith Stein's phenomenology of sensual and emotional empathy," *Phenomenology and the Cognitive Sciences* 17 (2018): 741–760.

walking through a shopping mall. They were first asked to register their experiences in a calculative and analytical manner by categorizing objects as spectators. In a second step they were asked to register these two different environments in an experiential way by noticing how they move, breathe, listen, and what they see and hear in a man made and in a more natural environment. This exercise made them aware of how a calculative, objectifying approach dominates the way we construe an environment and increasingly experience it as well. For comparative purposes it was therefore important to have them juxtapose the calculative approach with an experiential approach of sensing themselves in nature, being part of it, and living in interaction with the environment. Unsurprisingly, there was a great difference in the degree of sense of wellbeing in a shopping mall and in a park. The mall often felt stressful and the park as a green environment gave a greater sense of wellbeing and relaxation.

These two approaches, the calculative and the experiential, coincide with Eugene Gendlin's useful distinction between types of environment in his process philosophy.[11] Gendlin's main idea is that man-environment is one, a continuous interaction process. The calculative approach to the environment corresponds to his concept of Environment#1 which is the environment of the spectator who decides how the environment affects a living organism. The scientist and the hunter define the environmental factors in their own respective terms. Environment#2 is the environment that is identical with the organism's living process. Body and the environment are one living process as evident in the fact how air enters or water enters the body, is processed there and then leaves the body. Environment two is not a separable environment but the environment participating in the life process. Environment#2 corresponds to the experiential approach. As does Environmentt#3 as that which has been arranged by the body-environment#2 process, be it the spider's web, the beaver's tree, human artefacts, constructs and practices within them. Environment#3 can also correspond to posthumanist and neomaterialist conceptions of the environment as an interaction of the living, the human, the non-human, and the machine.[12]

On the basis of these distinction, the human body is not only the link to the living, man made and inorganic environment. The human body partakes in the environment as a living process. This approach transcends traditional dualistic divisions of nature and culture which have led ecological philosophers like Timothy Morton to abandon the concept of "nature" as part of the problem that has roots in a disconnection of the human from nature, emanating from objectifying and calculative thinking.[13] In the course of these developments, "nature" has in environmental philosophy come to be seen as a romanticized other to culture, and been put on a

---

[11] Eugene T. Gendlin, *A Process Model* (Chicago, IL: Northwestern Univeristy Press, 2017), Chapter 1.

[12] Karen Barad, *Meeting the Universe Halfway: Quantum Physics and the Entanglement of Matter and Meaning* (Durham: Duke University Press, 2007); Rosi Braidotti, *The Posthuman* (London: Polity, 2012).

[13] Timothy Morton, *Ecology without Nature* (Cambridge, MA.: Harvard University Press, 2009).

pedestal, alienating us from it and hence making us less attuned to the ecological way of thinking Morton calls for.

Nature, Environment, Life and the Body

Despite Morton's convincing criticism of the dualistic understanding of nature as an "other" opposite to the human (that more over supports an instrumentalizing approach to nature) the title of the course on Ethics of Nature has not been changed to Ethics of Environment. One of the main reason for holding on the concept of nature is that it is still the common, everyday term for living nature, but more importantly the concept of nature has a notion of life and living, natality and becoming in it that the concept of environment does not have. As part of nature, the body and the natural environment are an interaction, and the concept of nature also implies how the living is a continuous, never ending process of birth, change, end and, rebirth, also involving the force of the elements.[14]

The concept of "life" is in fact more apt to describe what is termed with nature but then again the concept itself is not even clear for biologists, making it hard to distinguish clearly between life and death. A notion of a living process is for the context of an ethics of nature preferable to a concept of life as such. The main reason for sticking with the concept of "nature" is however a more specific one, and it consists in directing attention to the first person perspective as the basis of embodied and elemental thinking. The body as a living process is what connects us to natural and man-made environments, and the body as living and thinking is our basis situation and grasp of the world like phenomenologists of the body from Simone de Beauvoir, Merleau-Ponty, and onwards have argued.

The body as situation adds something very important to prevalent understanding of how we grasp the world from Kant's epistemology to contemporary constructivist theories about how we acquire knowledge. For Kant, the world is disclosed to us via human cognition that combines sense perception and the ordering mind. According to views that take the Kantian paradigm in a more constructionist direction, the world does not imprint itself on the mind but we imprint ourselves on the world. The human mind constitutes the world and a computational model of the brain is in continuity with this model, with the development of AI based on translating the workings of the mind into machine intelligence and machine learning. From the perspective of other life-forms and non-human life, this marks the zenith of an anthropocentric and calculative model of human knowing that has been instrumental for scientific and technological advances.

There is however, a blind spot in such one-sided constructionist approaches because there is a forgetting of acounting for who asks these questions and how they

[14] Gudbjörg R. Johannesdottir and Sigridur Thorgeirsdottir, "Reclaiming Nature by Reclaiming the Body," *Balkan Journal of Philosophy* 8, no. 1 (2016): 39–48.

are asked on the basis of experiential and embodied thinking. Excluding the first-person embodied perspecive of the researcher from the reflection leads to a partial reflection, and the inquiry becomes disembodied and detached. In doing so, the scientist is excluded as a valuable source of investigation. Embracing the embodied turn allows for a deeper understanding of the subjective experience of knowledge acquisition. Acknowledging how indvididuals sense and experience their situation extends our understanding of the social situatedness of knowing. This first-person experiential, embodied, embedded knowing can provide insights into how individuals experiences motivate to change situations they perceive as repressive, restrictive or ecologically problematic.

From Disconnection to Connection: Lived Experience and Felt Sense

The embodied turn implies connecting with one's own lived experience, to connect with sensory levels of one's thinking and understanding, and also with levels of thought that are not yet cognitively captured or verbally or symbolically articulated. Claire Petitmengin has convincingly discussed how our capacity to connect with these levels of thought as lived inner experience is repressed in a culture that is driven by rationalistic and technological mastery. Or to put it more differently, such mastery and the undeniable benefits it has brought for wellbeing, life expectancy and comfort, has come at a cost of a disconnection with lived experience:

> this way of life and the ecological disaster it is bringing about are intimately linked to our blindness to that which is nevertheless closest to us: our own lived experience. Most of the time, we are cut off from ourselves, from what vibrates and lives within us, and this disconnection has catastrophic consequences in all areas of human existence. Retrieving contact with our experience is the precondition that would allow us to regain our lucidity, our dignity, and the courage to change our model of society.[15]

It is the felt dimension of experience that Petitmengin discusses here and that Gendlin has termed as the felt sense of an issue, condition, situation or phenomenon.[16] Felt sensing is a broader notion than sensual perception, that is perception that is restricted to the data given through the five senses (touch, hearing, sight, smell and taste). It is an embodied knowing on the basis of the knowledge experientially accumulated in bodies through living and learning. Gendlin therefore speaks of the primacy of the body as a broader source of information than that of perception in his discussion of Merleau-Ponty's phenomenology of perception (Gendlin 1992). Felt sensing can also not be reduced to the felt in the sense of feelings, emotions and

---

[15] Claire Petitmengin, "Anchoring in Lived Experience as an Act of Resistance," *Constructivist Foundatitons* 16, no. 2 (2021): 172.

[16] Eugene T. Gendlin 1966, "The discovery of felt meaning," in *Language and meaning*, *Papers from the ASCD Conference, The Curriculum Research Institute*, ed. J.B. McDonald and R.R. Leeper (Washington, DC: Association for Supervision and Curriculum Development): 45–62.

sensations. These main types of feelings are more specific than a felt sense, depending on how much they are cognitively loaded. An emotion like anger is for example based on a belief, a knowledge or an opinion about an issue that makes one feel angry. A feeling like shame that makes one blush is perhaps more evasive and diffuse and a less cognively laden feeling than anger. Gendlin dinstinguishes felt sense as a kind of philosophical sense of a whole of a situation:

> I distinguish emotion from "felt sense." A felt sense is a vague, implicitly complex, physical feeling that can come in your body in regard to any situation or any aspect of life. I will try to show why in psychotherapy we must focus on the felt sense. To do so seems unpromising at first. A felt sense seems vague and also much less intense than an emotion. We must surely welcome all emotions during therapy, especially blocked ones that have not been felt. But, the felt sense is needed to reach what gives rise to emotions, as well as for new steps.[17]

A felt sense is therefore something broader, a wholistic sense and partly similar to what Martin Heidegger conceptualized as a mood ("Stimmung") as a special category of emotions, a broad disposition that gives one a sense on feeling-being in a situation as a whole.[18] Whereas Heidegger discussed such emotions as opening up to a sense of the whole of one's existence, contemporary phenomenologists like Matthew Rattcliffe have come up with the term of existential feelings as a fundamental human affectivity.[19] Existential feelings are rooted in our embodied experience of being in the world and are not cognitive or intellectual states. It therefore takes an inward turn towards one's existential feeling or felt sense to connect with it and then attempt to articulate it. Environmental philosophers like Gernot Boehme talks about such a sense in an environment as an atmposphere and Timothy Morton discusses "ambience" as a state of "nondual awareness that collapses the subject-object division, upon which depends the aggressive territoralization that precipitates ecological destruction."[20] What Morton is in effect advocating is connecting with inner lived experience that is repressed in a culture that is still permeated by dualistic ways of thinking.

Exercises in Listening and Felt Sensing

In our course on Ethics of Nature students are introduced to embodied ways of thinking by beginning to learn practices of active listening. Students are given guidelines for active listening that distinguish the roles of a speaker and a listener. They realize that the role of the listener is far more complex than typically

---

[17] Eugene T. Gendlin, "On emotion in therapy," in *Emotion, psychotherapy and change*, ed. J.D. Safran and L.S. Greenberg (New York & London: Guilford): 255–279.

[18] Martin Heidegger. *Was ist Metaphysik?* Frankfurt am Main: Verlag Vittorio Klostermann, 1929.

[19] Matthew Ratcliffe, "The Phenomenology of Existential Feeling," in *Feelings of Being Alive*, ed. J. Fingerhut and S. Marienberg (Berlin/New York: de Grutyer, 2012).

[20] Timothy Morton, "Why Ambient Poetics," *The Wordsworth Circle* 33, no. 1 (2002): 52–56.

emphasized in a culture that prioritizes self-expression and speaking over listening. The listening exercise is structured so that a speaker talks about a given topic for a given amount of time. The role of the listener consists in giving the speaker a space to be able to speak and that means mereley giving silent signs that they are listening attentively. This means that the listener does not interfer, comment or ask questitons unless they do not hear or understand what the speaker says. After the given amount of time, the speaker stops. The listener has then a few minutes to reflect back to the speaker what they said by giving back the exact words or concepts and refrain from using their own. Therefore, a pivotal aspect of the listener's role is to adopt an attitude of refraining from judgement and interpretation.

What is perhaps most important in this exercise is that the listener learns to listen to themselves while listening to another person. The listener begins to notice how one tends to be a inattentive listener by being impatient and merely waiting for an opportunity to talk, using the other person's words as a stepping stone to formulate one's own ideas instead of truly attempting to hear and understand what the speaker is saying An exercise in active listening is thus in effect a first introduction to an inner listening, to an inward sensing. This exericise is not only important as a methood of conducting a dialogue more thoughtfully, but also a first step into felt sensing.

A main responsibility of philosophy teachers is to train students to think independently. Consequently, the practice of embodied and experiential listening has become a valueable and essential tool for strengthening students own thinking and more importantly increasing confidence in their own thought and reflections. Each person is situated uniquely in time, and space, i.e. in different contexts. Everyone has therefore a distinctive perspective and experience and potentially something unique to contribute to the world. The main task of these exercises is not only to allow students to connect with their own inner sense of an issue but also to enable them to articulate it verbally. There is often a meditative step involved in accessing inner experence or a felt sense for one needs to begin by calming down and clearing one's mind. The goal is to access deeper sensory layers of cognition and to search for the right words to express them so there is a fit beetween the inner sense and the word one finds for it.

Deborah Bird Rose is aware of how listening is important for increasing environmental awareness when she proposes that "listening, and more broadly, paying attention, should also be considered an actie verb; indeed in multispecies creature communites, it must be considered. To pay attention is to exercise intelligence, to know so as to be able to inter-act."[21]

With exercises adopted from microphenomenological interview methods, students are trained in using the knowledge gained from descriptions of their own experiences of environmental phenomena to reflect and formulate their own philosophical ideas and ethical positions. The different steps and moves of Gendlin's

[21] Deborah Bird Rose, "Val Plumwood's Philosophical Animism: Attentive Inter-Actions in the Sentient World," *Environmental Humanities* 3 (2013): 102.

methodology of Thinking at the Edge for scholarly and research practices are also introduced.[22] Gendlin's methodology consists in going back and forth between an experiential sense for an issue and conceptual, logical and argumentative aspects of it. One example of Thinking at the Edge is working with instances of experiences as a source for the students scholarly reflections of a particular topic or problem. In further steps, students learn to conceptualize their topic, formulate a key sentence, and construct their theory or hypothesis about their topic.[23] The results of such exercises are new insights into environmental issues and a deeper understanding of possibilities and limits of theories. The connection with students own experiences is the crucial factor for methodologies of elemental and embodied thinking. Their first person perspective does not remain merely a subjective account because reflecting their felt sense of an experience allows students to formulate something general on the basis of it. The theoretical abstractions that result hence are grounded in students as embodied and elemental beings.

Embodied, Elemental Thinking in Research

Students in Ethics of Nature are not only trained in felt sensing of their own experiences of environments and environment related phenomena. These methods can also be applied for research. Let me give an example. Several years ago there was an earthquake spell that lasted three weeks preceding a volcanic erupotion in an unpopulated area outside Reykjavik. There were numerous earthquakes every day that were acutely felt but did not cause destruction. Most people adopted the common attitude in Iceland (a volcanic island placed on the Mid-Atlantic ridge that separates the Eurasian and North American tectonic plates), to let the earthquakes pass by without being startled by them. There is however a group of people who get very nervous when the earth begins to tremble and to shake. I conducted microphenomenological interviews with individuals who experienced stress during the earthquakes to gain insights into their experiences. A sense of disconnection was a common feature of their experiences of the earthquakes. They reported feeling powerless and subjected to external forces that vastly exceed human scale, leading to an overpowering feeling of lack of control.

I interviewed one person twice. This person described how nervousness got the better of her, and how she ended up crying at the dinner table as she could not take "anymore of those quakes." The next day, I interviewed her about an experience of an earthquake that occurred earlier that morning. When the earthquake began, something told her to pause and tune into her senses. As she reflected on her experience, she described feeling a deep connection between the trembling earth and her

[22] Eugene T. Gendlin, "Introduction to Thinking at the Edge," The Folio 19 no. 1 (2004): 1–8. https://www.focusing.org/gendlin/docs/gol_2160.html.

[23] Donata Schoeller and Sigridur Thorgeirsdottir, "Embodied Critical Thinking: the Experiential Turn and its Transformative Aspects," *philoSOPHIA* 9 no. 1 (2019): 92–109.

own body, particularly her womb. It was as if there was a cord linking her womb to the earth. This cord-like connection made her feel calmer and enabled her to endure the earthquake without getting as distraught as she had in previous days. Perhaps this was an unexpected therapeutic effect of the interviews or just this person learning to deal better with the stress during earthquakes. In a sense she was experiencing a crux, being part of a shaking earth that made her disconnect from herself and the earth in a paralysis or flight reaction to it. This does not mean that she overcame her fear of the earthquakes by connecting to herself and the earth but it meant that she had a greater spectrum of reacting to the earthquakes, allowing her to sense into this earthly movement rather than being overridden by a sentiment of wanting to flee it or to become paralysed and feeling stuck.

It would require a separate paper to argue and explain how connecting with one's felt sense of an environmental situatiton can inspire environmental action. However, the primary focus of this paper is to argue that connecting with this implicit sense is an important prerequisit for discovering one's direction and find ethical orientation in environmental matters, which can ultimately lead to self-transformation, change and action. In order to understand embodied and elemental thinking as a practice it is however not enough to read about it. One has to try it out to get the gist of it.

Bibliograpy

Barad, Karen. 2007. *Meeting the Universe Halfway: Quantum Physics and the Entanglement of Matter and Meaning*. Duke University Press.

Bendell, Jem. 2020. *Deep Adaption: A Map for Navigating Climate Tragedy*. Available in different versions here: https://lifeworth.com/deepadaptation.pdf.

Braidotti, Rosi. 2012. *The Posthuman*. London: Polity.

Gendlin, Eugene T. 1966. The discovery of felt meaning. In *Language and meaning, edited by* J.B. McDonald & R.R. Leeper. Papers from the ASCD Conference, The Curriculum Research Institute (Nov. 21–24, 1964 & March 20–23, 1965), 45-62. Washington, DC: Association for Supervision and Curriculum Development. http://previous.focusing.org/gendlin/docs/gol_2039.html.

———. 1991a. On emotion in therapy. In *Emotion, psychotherapy and change*, ed. J.D. Safran and L.S. Greenberg, 255–279. New York/London: Guilford. http://previous.focusing.org/gendlin/docs/gol_2068.html.

———. 1991b. The primacy of the body, not the primacy of perception. *Man and World 25* (3–4): 341–353. http://previous.focusing.org/gendlin/docs/gol_2220.html.

———. 2004. Introduction to Thinking at the Edge. *The Folio* 19 (1): 1–8. https://www.focusing.org/gendlin/docs/gol_2160.html.

———. 2017. *A Process Model*. Chicago: Northwestern University Press.

Haraway, Donna. 2016. *Staying with the Trouble: Making Kin with the Chthulucene*. Durham: Duke University Press.

Heidegger, Martin. 1929. *Was ist Metaphysik?* Frankfurt am Main: Verlag Vittorio Klostermann.

Irigaray, Luce, and Michael Marder. 2016. *Through Vegetal Being: Two Philosophical Perspectives*. New York: Columbia University Press.

Johannesdottir, Gudbjorg R., and Sigridur Thorgeirsdottir. 2016. Reclaiming Nature by Reclaiming the Body. *Balkan Journal of Philosophy* 8 (1): 39–48.

Lorde, Audre. 2018. *The Master's Tools will never Dismantle the Master's House*. New York: Penguin.
Malm, Andreas. 2021. *How to Blow up a Pipeline: Learning to Fight in a World on Fire*. London: Verso.
Marder, Michael. 2013. *Plant-Thinking. A Philosophy of Vegetal Life*. New York: Columbia Univeristy Press.
Freya Matthews. 2018. *We've Had Forty Years of Environmental Ethics – And the World is Getting Worse*. https://www.abc.net.au/religion/weve-had-forty-years-of-environmental-ethics-and-the-world-is-/10214334.
Morton, Timothy. 2002. Why Ambient Poetics? *The Wordsworth Circle* 33 (1): 52–56.
———. 2009. *Ecology without Nature: Rethinking Environmental Aesthetics*. Cambridge, MA: Harvard University Press.
———. 2013. *Hyperobjects: Philosophy and Ecology after the End of the World*. Minneapolis: University of Minnesota Press.
Nussbaum, Martha. 2023. *Justice for Animals: Our Collective Responsibility*. New York: Simon & Schuster.
Petitmengin, Claire. 2021. Anchoring in Lived Experience as an Act of Resistance. *Constructivist Foundatitons* 16 (2): 172–181. https://constructivist.info/16/2/172.
Ratcliffe, Matthew. 2012. The Phenomenology of Existential Feeling. In *Feelings of Being Alive*, ed. J. Fingerhut and S. Marienberg. Berlin/New York: de Grutyer.
Regan, Tom. 2000. *Defending Animal Rights*. Champaign: University of Illinois Press.
Rose, Deborah Bird. 2013. Val Plumwood's Philosophical Animism: Attentive Inter-Actions in the Sentient World. *Environmental Humanities* 3: 93–109.
Schoeller, Donata, and Sigridur Thorgeirsdottir. 2019. Embodied Critical Thinking: the Experiential Turn and its Transformative Aspects. *philoSOPHIA* 9 (1): 92–109.
Singer, Peter. 1975. *Animal Liberation: A New Etthics for our Treatment of Animals*. New York: Random House.
Stein, Edith. 1989. *On the Problem of Empathy, The Collected Works of Edith Stein, Vol. 3*. Trans. Waltraut Stein. Washington, DC: ICS Publications.
Sveneaus, Fredrik. 2018. Edith Stein's phenomenology of sensual and emotional empathy. *Phenomenology and the Cognitive Sciences* 17: 741–760.

Sigridur Thorgeirsdottir is a professor of philosophy at the University of Iceland. She studied philosophy in Boston and Berlin. She has worked at the universities of Rostock, Helsinki and Iceland. She has published books on Nietzsche, Beauvoir, Arendt, feminist philosophy, and philosophy of the body. Sigridur recently co-edited a collection of essays on methodologies of the study of women in the history of philosophy, and has published three calendars with short entries on women philosophers from the beginnings of philosophy to the present (one of them for the World Congress of Philosophy in 2018 in Beijing as chair of the committee for gender issues of FISP). She is one of the leading investigators of the international research project Embodied Critical Thinking (www.ect.hi.is) and of "Training Embodied Critical Thinking", an Erasmus+ funded training program for methodologies of embodied critical thinking. A book on the research of the project is now coming out, Practicing Embodied Thinking in Research and Learning, co-ed. with Donata Schoeleler and Greg Walkerden (London: Routledge, 2024).

Index

L. Škof et al. (eds.), *Elemental-Embodied Thinking for a New Era*, Sophia Studies in Cross-cultural Philosophy of Traditions and Cultures 42,
https://doi.org/10.1007/978-3-031-42119-8

Printed in the USA
CPSIA information can be obtained
at www.ICGtesting.com
LVHW021545190724
785981LV00003B/134